I0475377

Table of Contents

1

1 Introduction

称之为原螺旋的统一场论回答了物理最基本的问题。例如，正负电荷，量子单位，电子的稳定性及电子所带单位电荷。它为物理学提供更好的基石，实现物理的完整无误，本文提出了新理论，给物理最基本的问题作出了圆满的解释。相对论研究时间与空间的关系。统一场论加上能量的概念，并且把能量与时空对等。而螺旋运动则把三种概念归结为一个可以想见的模型。简单的螺旋，可解释强力，弱力，电磁力及重力等四种基本力。本文将所有物理定律由新的理论一一推导出来。它为今后的物理学乃至科学的发展提供了新的思想和新的方法。

The Alpha Torque is a unified fields theory that attempts to answer fundamental physics questions, such as why positive and negative charges exist, why quantum numbers exist, and why the electron is stable and has unit charge. It provides a better foundation, improves the accuracy and completeness of Physics theories and answers fundamental questions. The Theory of Relativity studies the relationship between space and time. The unified fields theory adds the concept of energy and considers that space, time and energy are equally important. The Torque movements unify the three concepts into one visible model. The simple Torque model can explain four fundamental interactions, strong interactions, weak interactions, electromagnetic forces, and gravity. Any of the exiting theories can be derived from the new theory. It provides new concepts and new methodologies for the development of future physics and science.

为了便于理解，复杂的数学推演步骤放在数学分析一章中。

In order to make the book understancable, the complex analytical steps are moved to Mathematical Analysis.

本书采用中英文对照。因为我的同胞们想看中文书。他们中有的人和我一样，看中文比看英文容易多了。

This book has Chinese and English. My fellow Chinese want to read a Chinese book. Some of them just like me; it is much easy for them to read Chinese than to read English.

子曰，"朝闻道昔死可也。"

Confucius said, "If a man hears the Way in the morning, he may die in the evening without regret."

2 Torque and Universe

2.1 Paradox of Relativity

相对论是基于"单程光速恒定"公设：

Relativity is based on the "constant one-way speed of light" postulate:

无论观察者的相对运动或光源的运动如何，光在真空中的速度对所有观察者而言都是相同的。

The speed of light in a vacuum is the same for all observers, regardless of their relative motion or of the motion of the source of the light.

在以上宣称中，条件"所有观察者"意味着有许多不同速度的观察者，而最简单的情况下有两名观察者。条件，"不管光源的运动如何"意味着有许多光源，以不同的速度移动，而最简单的情况下，有两个光源。它可以进一步简化为两种情况：

In the above claim, the condition "*all observers*" implies that there are many observers with different speeds and the simplest case has two observers. The condition, "*regardless of the motion of the source of the light*" implies that there are many light sources which move at different speeds and the simplest case has two light sources. It can be further simplified to two cases:

1 观察者 A（x，y，z）的与光源的速度移动相同，另一观察者 B（x'，y'，z'）向光源 L 方向移动的速度为 v 而 v> 0。

2 光源为 L'而 L'在 B 中静止不动，其它条件与上述情况相同。

1. One of the observers A (x, y, z) is moving at same speed with the light source, another B (x', y', z') is moving toward the light source L at speed v and v > 0.

2. The same as the above, except light source is L' and L' is not moving in B.

In case 1:

$$t' = d(t - xv/c^2)$$

$$x' = d(x - vt)$$

$$y' = y$$

$$z' = z$$

In case 2:

$$t' = d'(t + xv/c^2)$$

$$x' = d'(x - vt)$$

$$y' = y$$

$$z' = z$$

上面的方程得出这样的结论：

The above equations conclude that:

d = d' = 1

v = 0

(x, y, z, t) = (x', y', z', t')

上述结果不合乎逻辑，因为，

The above results are not logical because,

(x, y, z, t) != (x', y', z', t')

v != 0

相对论公设的悖论把物理学带回到现代物理学在一个世纪前的起点，以太理论。同相对论相比，以太理论是同样糟糕或更差；因此，本文提出了一种新的转矩理论（统一场理论），取代相对论和以太理论。

The paradox of relativity postulate brings us back to the theory of aether which is the starting point of modem Physics a century back. As compared to the theory of relativity, the aether theory is equally bad if not worse; therefore, this paper proposes a new Torque Theory (Unified Fields Theory) that replaces both the Theory of Relativity and Theory of Aether.

2.2   Universal Torque

Arbitrary 3D Movements

真的三维运动是方向不断改变，前后关联的捻动。有两种可能的捻动方式，Z 捻动和 S 捻动。 除了捻动，还有直线和圆周运动。但只有捻动在三个维度中都有运动。

The directions of real arbitrary 3D movements are changing and correlated. There are two possible movements: S twist and Z twist. Besides twist movements there are straight lines and circular movements as well, but only twists have movements in each of the three dimensions.

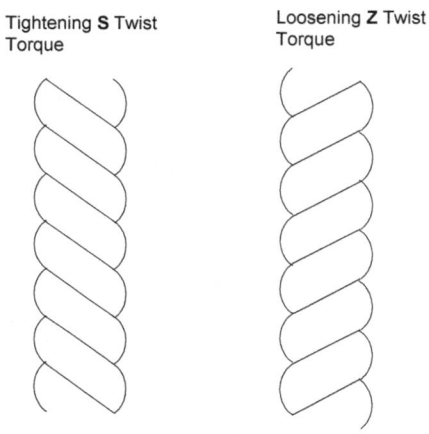

Figure 2-1

捻动在这儿称之为螺旋。上面左图为紧螺旋，右图为松螺旋。松与紧是指两个不同螺旋方向。在日常生活中，螺旋处处可见。例如，把盖子拧上罐口是紧螺旋运动。拧开一扇门的把手则是一个松螺旋运动。

The Twist Torque movements are called Torque movements. The diagram on the left is a tightening torque, while the one on the right is a loosening torque. "Tightening" and "Loosening" are the two different directions of Torques. The Torque movement appears in everyday life. For example, the act of screwing a lid onto a jar is a "Tightening" Torque. When opening a door with a knob, a Loosening Torque is taking place.

The Law of the Torque

螺旋运动是空间，时间和能量的存在方式。螺旋运动的速度是光速并有固定的平均波长。

The space-time-energy exists as Torque movements at speed of light with a fixed average wavelength.

2.3   Torque String and Torque Grid

在三维坐标系统中选择好方向，用螺旋运动的平均波长为单位便可得到一个立方体。这样便形成螺旋网格系统：

One can choose three directions in 3D coordinator system and use average wavelength to form a cube. A Torque Grid system emerges:

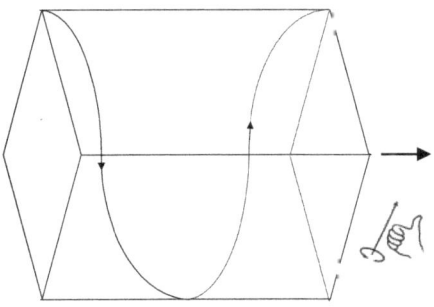

Figure 2-2

上图是一个单位波长的螺旋的网格。为了便于记忆，本理论把宇宙的螺旋设定为遵循右手法则的 Z 松螺旋。因为左右相冲，左手螺旋不能是本宇宙的主要螺旋运动方向。

The figure above has a single Torque Grid with a single Torque cycle. To make it easier to remember the movement of the Torque String, this theory models the String movement as right-handed Z loosening Twist movement. Left-handed Torque Movement is not the main movement of the universe, since it conflicts with the right-handed movements.

The Laws of Torque Grid

13

螺旋线以光速周期式以右手法则运动，形成螺旋的网格，代表时空和能量。

The Torque Strings moves in cycles at light speed, follows Right Hand rule, and forms Torque Cubes and represents space-time-energy.

2.4 Energy Torque Interaction

The Law of Energy-Torque Interaction

螺旋的网格受能量的影响而增大。而局部螺旋的网格大小的起伏变化不影响整个宇宙的平均网格的大小。

The energy enlarges the Grids. The local energy level fluctuation has no impact to the average Grid size in the universe.

当受能量 E 影响而波长增大的螺旋线向外运行时，与正常螺旋线的相位差逐步加大。在距离 L 处，相位差增加到一个螺旋网格的大小，变形的螺旋线与正常螺旋线同步。L 是能量 E 的波长。能量越大，形变越大，波长 L 越小，频率 v 越高：

When the enlarged Strings by energy E with different cycle go outward, the cycle phase difference between the distorted Strings and normal Strings is increasing. When the phase difference increases to the size of a Torque cube at a distance of L, the distorted string phase synchronizes with the normal Grid again. L is the wavelength of the energy of E. The more of energy E, the bigger the Torque Grid distortion, the smaller of L and the higher frequency v:

$$E = hv \qquad (2\text{-}1)$$

以上是普朗克公式。

The above equation is the Planck equation.

普朗克公式已为实研所证实。它为本理论提供了实研证明。螺旋网格理论为普朗克公式作了理论解释。

The Planck equation is experimentally proven. It provides an experimental proof for this theory. The Torque Grid theory provides theoretical explanations of the Planck equation.

2.5 Gravity Field

当质量为 m 的物体进入质量为 M 的物体影响而形变的空间时，因两个物体的参照螺旋网格的增大而产生了多余能量。这就是引力的来源(详细过程请看数学分析一章)：

When mass m comes to distorted (enlarged) space, the referenced Grid sizes for both bodies are increased. Thus, there is left over energy. This is the origin of the gravity forces (See Mathematical Analysis for further detail):

$$f = \frac{GmM}{R^2}$$

(2-2)

2.6 Torque Grid Size

螺旋网格的长度是由引力常数 G，普朗克常数 h 和光速 c 所决定：

Torque grid's size can be determined by the gravity constant G, plank constant h and speed of light c:

$$D = \sqrt{\frac{Gh}{\pi * c^3}} = 2.2856509 * 10^{-35} m$$

(2-3)

弦论中的单位空间长度是普朗克长度 l_p 。下面是螺旋网格的长度与普朗克长度的关系：

The unit of space in String Theory is Planck length l_p . The following is the relationship between the size of Torque Grid and Planck Length:

$$D = \sqrt{2} l_p$$

2.7   Electron Stability

电子稳定的原因是沿电子壳的总扭曲是螺旋网格的长度。

An electron is stable because total distortion along the shell of electron is the length of the Grid.

换言之，电子因在其壳上和螺旋网格共振而稳定。一个小于光的速度移动的粒子称为慢粒子。螺旋网格的大小和属性决定了电子质量和大小。它是最小最稳定的的慢粒子。因而电子不能衰退为更小的慢粒子。

Electrons have resonance points on their shells with Torque Grid. A particle moving at less than light speed is called a slow particle. The size and property of the grid determine the mass and size of the electron. It is the smallest and most stable slow particle. Therefore electron cannot decay into smaller slow particles.

2.8   Torque Grids Unify Space-Time-Energy

螺旋网格具时间，空间和能量的特征。螺旋网格有固定的位置作空间的参照系，而螺旋网格的大小是空间的单位。螺旋弦有如同时间的单方向性，而螺旋弦的固定的周期是时间的单位。螺旋网格的扭曲是能量的特征。电子因在其壳上同螺旋网格共振而稳定。电子所含能量是单位能量。这样，螺旋网格把时间，空间和能量统一为一个最简的形式。

The Torque Grid has the characteristics of space, time and energy. The Torque Grid's fixed location is the reference for the space and size is unit of space. The Torque String has one directional time characteristic and its fixed cycle time is the unit of time.   The Torque Grid twisting distortion is character of energy and electron has resonance points on its shell with Torque Grid. An electron's energy is the unit of energy. Therefore, Torque Grid unifies the space, time and energy in the simplest form.

3   Particle

如前章所述，电子是最稳定的粒子。它带有单元电荷，永远不会衰竭，是一切基本粒子的基石。

The electron is the most stable particle as mentioned in the previous chapter. It has one single charge, and it never decays. It is the building block of fundamental particles.

3.1 Torque Waves

粒子内螺旋波带有单位电荷并以电子的频率为单位。其频率遵循斐波纳契数列：

A Torque wave inside a particle is a unit charged wave and the unit of its frequency is the frequency of the electron. The frequencies follow Fibonacci series:

2, 3, 5, 8, 13, 21, 34, 58, 89, 144 …

能量方程：

The energy equation:

$$E_{total} = Sv_1v_2...v_n \qquad (3\text{-}1)$$

S 为起始能量单位。它可为 1 或 2。

S is the base power. It can be 1 or 2.

3.2 Strong Forces

当粒子的质量是电子的质量的 137 倍时，它粒子的壳上产生一个螺旋网格长度形变。同电子相比，能量是 137 倍。因此，相互作用力比电场力量力大 137 倍。

When the particle's mass is 137 times electron's mass, it creates one Grid length distortion on the shell of the particle. Comparing to electron, the energy is 137 times greater. Therefore, the interactive force is 137 times greater than electronic forces.

3.3 Wave Formula of Strong Forces

根据以下斐波那契数列，可发现 137 这个数值有某些特殊的含意：

The number 137 has some special meanings based on the following Fibonacci series:

3, 5, 8

其相互作用的方式可用一个简单的质量算式表示：

The way of interactions can be expressed as a simple mass formula:

3*5*8 + 3*5 + 2 = 137

3.4 Prediction of Weak Forces

电子外壳上的的普朗克形变是：（1/137）螺旋网格大小。力是单位电荷力的（1/137）。总能量是单位电荷能量的(1/2)（1/137）*（1/137）。

The Planck distortion on the shell of an electron is (1/137) Grid size and the force is (1/137) of electron force. The total energy is (1/2)(1/137)*(1/137) of electron energy.

3.5 Energy Distribution in Particles

以下是电子的能量分布涵数（螺旋网格大小为单位）：

The following is the electron energy distribution function (Torque Grid size is the unit):

$$1/(47.5n) \tag{3-2}$$

4 Particle Formula

4.1 Basic Structures

电子是基本粒子。它有单元电荷，在其外壳上有一个螺旋网格的形变，而且永远不会衰变。比电子质量大的粒子都是以电子质量为单位的能量波的共振得以稳定而形成的。每个基本波带有一个单元电荷。

The electron is a fundamental particle. It has single charge and single Torque Grid distortion and it never decays. The particles more massive than the electron are formed as result of the resonance of energy waves with the electron mass as their unit. Each wave unit has a single unit charge.

粒子的最基本结构是带单位电荷的波。其能量值为以电子质量为单位的斐波纳契数列：

The particle's basic structures are unit charged wave structures. The energy values follow the Fibonacci series (electron mass is the unit):

1, 2, 3, 5, 8, 13, 21, 34, 58, 89, 144 …

or:

2, 2, 4, 6, 10, 16, 26, …

本章将假设：

In the rest of the chapter, assume:

A = 2*3*5

A^2 = （2*3*5）*（2*3*5）

B = 2*2*4

$B^2 = (2*2*4) * (2*2*4)$

4.2 Leptons

μ 轻子

Muon

2*(2*2*4*6)+3.0305*4.0305 + 2.55

τ 轻子

Tauon

2*9*10*19+ + 2*19 +19

4.3 Hadrons

质子:

Proton:

2* A^2+A +2*(3 .076335)

中子:

Neutron:

2* A^2+A+2.03034978*3.03034978+ 2.5309

ρ 介子

Rho meson

A^2+ 2*B^2 + 2*A + 2*B + 2* 4.47 +4.47

π 介子

Pion meson

$B^2+(2*2*4.28)$

K 介子

Kaon meson

$A^2+2*A+6.082$

Δ 重子

Delta Baryon

$2*A^2+2*A+2*B^2+2*B+3 + 3$

Λ 重子

Lambda Baryon

$2*A^2+4*A+B^2+1.825+1.825+1.825+1.825$

Σ 重子

Sigma Baryon

$2*A^2+2*B^2+ B^2 + 5.94$

$Xi^0= 2*A^2+3*B^2+2.56 + 2.56$

Ω 重子（带一个负电荷）

Omega Baryon⁻

$3* A^2+2*B^2+A+ B+2*4.97 + 4.97$

Ω 重子（零电荷）

Omega Baryon⁰

$4* A^2+6* B^2+ 3*A+ 3*B+2*2.45$

J/ψ 介子

J/psi meson[0]

$6*A^2 +2*B^2 + 3*A+3*B+2*5.24$

自反粒子

Self Anti Particles

η 介子

Eta meson[0]

$4*B^2+ 2*B + 8 + 8$

η′介子

Eta prime meson[0] $= 2*A^2+ 2*A+ 7 + 7$

Φ 介子

Phi meson[0] $= 2*A^2+ 6*A + 6*2.5$

4.4 The Meaning of the Particle Formula

含有 A^2 或 B^2 的粒子称之为强子。

The particle with A^2 and B^2 components are Hadrons.

在强子中的 A^2 和 B^2 决定了粒子的主要特性。重子和介子没有分别。可把重子划分为仅含有 A^2 仅含有 B^2 或同时含有 A^2 和 B^2 的三种类别。

A^2 and B^2 in Hadrons determine the characteristics of the particle. There is no difference between Mesons and Baryons. The Hadrons can be further categorized into the three categories: A^2 only, B^2 only or with both A^2 and B^2.

自反粒子分析：

Self Anti Particles formula analysis:

η 介子

Eta meson^0

$4*B^2 + 2*B + 8 + 8$

η′ 介子

Eta prime meson^0

$2*A^2 + 2*A + 7 + 7$

Φ 介子

Phi meson^0

$2*A^2 + 6*A + 6*2.5$

自反粒子不同时含有 A^2 和 B^2 。在粒子中的基本波成对出现。

The Self Anti Particles cannot have both A^2 and B^2. The basic waves are in pairs.

质子是除电子外的所有其他的粒子中最稳定的粒子。质子的结构为 $2* A^2 + A + 2*(3 .076335)$。它可以简化为 $2* A^2 + A + 2*3$。质子内的波有多个共振关系，A 与 A^2，A^2 与 A^2，$2*3$ 与 A 的共振，A^2 与 A^2 有形式对称十字结构，其垂直轴上有 A，从而在整体结构是对称的，等等。在质子的基本波间没有不和谐的波。不和谐是湮灭的根源，没有不和谐的粒子就不会自动湮灭。质子的稳定的根源全在于对称与和谐。

The proton is most stable particle of all particles, other than the electron. The structure of proton is $2* A^2 + A + 2*(3 .076335)$. It can be simplified to $2* A^2 + A + 2*3$. The

23

waves inside Protons are in resonance in many ways, A $\leftrightarrow$ A^2 resonance, 2*3 $\leftrightarrow$ A resonance, A^2 $\leftrightarrow$ A^2 resonance, A^2 $\leftrightarrow$ A^2 form symmetrical structure with A on its vertical central axis, etc. The whole structure is symmetrical. There is no dissonance found in the Proton formula. Dissonances among the basic waves are the reason behind the annihilation of the particles. In the other words, the particle does not annihilate without dissonance. The reasons for Proton's stability are symmetrical and resonance.

5   Structure of the Nucleus

5.1   Topology of Proton

质子的质量式为：2* A^2+A +2*(3 .076335。在三维坐标系统中，x → A^2，y → A^2，z → A 形成八面体结构。

In Proton's formula 2* A^2+A +2*(3 .076335), in 3D coordinator system, x → A^2, y → A^2, z → A and form the octahedron structure.

5.2   Topology of the Nucleus

核是个八面体与质子同形。以一个或两个基础正方形为开始向两个轴心方向对称堆积质子。可能的基础结构是，2×2，4×4，8×8，16×16，…

Nuclei have a shape of an octahedron to match the shape of the Proton. The structure starts with one or two base squares and accumulates smaller squares along the axis of the base squares in both directions. The possible base structures are, 2*2, 4*4, 8*8, 16*16, …

Formula	Symmetric	Element	Stable Isotopes
4*4 + 2*(3*3) + 2*(2*2) + 2*1	Yes	Ru (44)	7
4*4 + 2*(3*3) + 2*(2*2)	Yes	Mo (42)	4
4*4 + 2*(3*3) + 2*(2*2) + 1	No	Tc (43)	0

2*(4*4) + 2*(3*3) + 2*(2*2) + 2*2	Yes	Sm (62)	4
2*(4*4) + 2*(3*3) + 2*(2*2) + 2*1 +1	No	Pm (61)	0
4*4 + 2*(3*3) + 2*(2*2) + 8*6 +2*1	Yes	U (92)	0, but 235U has long half life

6 Meaning of Relativity

本理论可以推导出相对论方程。

The Torque theory can derive Special Relativity equations.

The Law of the References

在惯性参考系里和在螺旋网格绝对参照系的物理实验应当得到相同的结果。

The physics experiment in the inert reference system yields the same results as the Torque Grid absolute reference system.

当移动参考系在螺旋网格参照系中以速度 v 移动，空间/时间膨胀 1/d 发生在所有方向上。

When the moving reference moves at speed v in Torque reference, the space/time expansion occurs in all directions with same factor 1/d.

$$d = \sqrt{1 - \frac{v^2}{c^2}}$$

螺旋网格绝对参照系可得出相对论的所有公式。例如：

We can get all relativity's formula using the absolute Torque reference. i.e.:

$$E_{moving} = \frac{E_{torque}}{\sqrt{1 - \frac{v^2}{c^2}}} \tag{6-1}$$

$$E = mC^2 \tag{6-2}$$

27

7 Firmament of the Universe

7.1 Dark Energy

能量不是以质子，原子等可见的形式存在，就是以黑能量的形式存在。与可见能量不同，除电子以外的黑能量可均匀分布。当引力场足够大时，黑能量分布如下：

When the energy is not in form of protons, atoms and other forms of "visible" forms, it is in the form of Dark Energy. Unlike visible energy, the density of Dark Energy other than electrons can be smoothly distributed. The Dark Energy has following mass distribution when the galaxy gravity is large enough:

$$M = Kgr \tag{7-1}$$

g=2.63912*10^{20} kg/m

7.2 Galaxy Rotational Curve

公式(7-1)决定了每个星系都有以下固定的恒星绕行速度：

The equation (7-1) decides that a galaxy has a constant rotation speed as follow:

$$V = K^{1/2} *131.5 \text{ km/s} \tag{7-2}$$

K 可有以下值以满足与引力速度的共振：

The value of k needs to have proper resonance with gravity speed 131.5 km/s, where,

K = 1/8, 1/7, 1/6, 1/5, ¼, 1/3, ½, 1, 4/3, 3/2, 5/3, 2, 7/3, 5/2,8/3, 3 …

7.3 Arms and Bars of Galaxies

当两个星系相互靠近，两个星系之间的圆周运动相互抵消而速度大小保持一致。这样，黑能量的移动方向变为沿着两个星系之间的方向而形成一个"棒"的结构。运动贯性把"棒"延伸到星系的另一侧，形成在两个星系外面的星系的"旋臂"。即使两个星系之间的合并完成后，棒和旋臂结构在新的星系中心将保持一段时间。

When two galaxies with same rotation direction meet, the circular movements between two galaxies are cancelled out, but the speed remains the same. The dark energy moving direction will be along the two galaxies and form a "bar" structure in between. The moving momentum extends the "bar" to the other side of the galaxy and completes the "arm" structure. Even after completion of merging between two galaxies, the bar and arm structure will remain for a while in the new galaxy bulge.

7.4   Black Hole

根据能量守恒定律，损失的质量不能大于原来的质量，或（GMm / R）/ mC2 <1。因此，黑洞理论就不成立，或者说，重力不能捕获光子。因此，宇宙中的恒星虽然会消逝，但恒星里的物质却要被新的恒星回收；宇宙是永恒的。另外，热力学理论只可说明局部的现象，不能推广到整个宇宙。否则，就会得出宇宙正在变老最终走向死亡的错误的结论。

According to the law of energy conservation, the lost mass cannot be greater than the original mass, or $(GMm/R)/mC^2 < 1$. Therefore, gravity cannot trap a photon. Stars in universe can be recycled, but cannot be "dead"; the universe is timeless. The thermal dynamic theory does not work at universe level, since it predicts that the universe is succumb to heat death.

7.5   Non-Scattering Electron Photon Interaction

光子和电子的相互作用有一个新的模型。在相互作用中，光子的方向不变。在大多数情况下，光子在相互作用过程中发生红移。

A new model of the interaction between photons and electrons is proposed. During this interaction, the photon's moving direction does not change. In most cases, photon is red-shifted during the interaction.

这个理论给红移现象提供解释，而挪去了大爆炸理论的基础。为即螺旋网格理论提供了实验基础。

This theory removes the bases of the Big Bang theory by giving red-shifting proper explanation. It provides experimental foundations for Torque theory.

7.6 Prediction of Universe Boundary

宇宙在 2.49 * 10^{26} 米的距离，达到一个概念极限，即宇宙螺旋网格。

At the distance of 2.49 * 10^{26} m, the universe reaches to a conceptual limit, a Universe Torque Grid.

每个宇宙都是一个螺旋网格。尽管宇宙螺旋网格巨大，但它是更高层螺旋网格中的最小单位。

Each universe is a Torque Grid. Even though the universe Grid is big, but it is the smallest unit in the next Torque Grids' hierarchy.

7.7 Prediction of Torque Grids' Hierarchy

螺旋网格的层次构造可解释宇宙从低级螺旋网格到高级螺旋网格的大结构. 我们的宇宙是一个被称之为"宇宙网格"的螺旋网格。每层螺旋网格结构同样有空间，时间，能量的概念，遵循同样的和物理定律。因此，网格是相似的。像我们的宇宙一样，更高层次的宇宙和较低层次宇宙都有的星系和恒星。我们的宇宙会继续改变，但它永远不会死，也不会因改变它的年龄而变老。宇宙螺旋网格就是宇宙的穹苍。

The hierarchy of the Torque Grid explains the grand structure of the universe, from Torque Grids to Universe Grid. The universe is a Grid in the Torque hierarchy called universal Grid. Each hierarchy shares the same concept of space, time, energy and physics' laws. Therefore, Grids are similar. The higher universe and lower universe can have Galaxies and stars like our universe does. Our universe will continue to change, but it will never die, nor change its age. The Universe Grid is the firmament of universe.

7.8 Detailed Analysis

Hierarchy Ratio Constant

螺旋网格的边长 D 的 N 倍是宇宙网格大小。D*N*N 则是比本宇宙更高一级的宇宙的大小。同样的原因，螺旋网格长度除以 N 就得到更低一级的螺旋网格的长度。

The Torque Grid size D times N is the universe Grid size. Then, universe Grid size times N is the Grid size of the next level Grid above the universe. For the same reason, the Torque Grid size divided by N is the size of Torque Grid one level lower than the Torque Grid.

$N = 6.14631 * 10^{60}$

请注意，上述结果是由目前已知的宇宙的密度计算出的。

Please note that the above result is based on the density of universe known today.

Universal Grid in the Vacuum

可见的宇宙在更高一级螺旋网格最可能是个真空中的一个小小的螺旋网格。

The visible universe can be a little Grid in the vacuum at the higher Torque Grid hierarchy.

Alternating Torque Directions in Hierarchy

较低或更高层次的网格和当前网格右手螺旋相比，螺旋方向相反而以左手螺旋方向运动。上层网格的能量降低了当前网格整体的能量密度，放大网格的大小。

The lower hierarchy Grids and higher hierarchy Grids have opposite Torque movements, or left-handed movement, as compared to the current right-handed movement universal movements. The upper level Grid's energy reduces the overall energy density at the lower level and enlarges the size of Grid.

Static Average Grid Size

引力对螺旋网格的增大仅改变局部螺旋网格大小。在整个宇宙中，平均螺旋网格大小不变。局部的螺旋网格大小波动并没有影响到整个宇宙，因为两个层次螺旋网格的边长比率是常数。 在引力作用下，周围区域的螺旋网格增大，而导致宇宙其余区域的螺旋网格缩小。增大和缩小相互抵消而整个宇宙的平均螺旋网格大小保持不变。

The gravitational enlargement of the Grids only impacts the local Grid size. At the universal Grid level, the Grid size does not change. The local level Grid size fluctuation has no impact to the universal Grid since the Grid ratio N between two hierarchy levels depends on energy density is a constant. The gravitational distortion of the Grids causes the surrounding the energy source to enlarge and causes shrinkage in the rest of the areas in the universal Grid. The enlargement and shrinkage cancel out each other at the universal Grid level and average Grid size stays static.

8 Particle Waves

粒子能量在其周围产生波状螺旋网格形变如下：

The energy of a particle creates Torque Grids' distortion in wave form around the particle as follows:

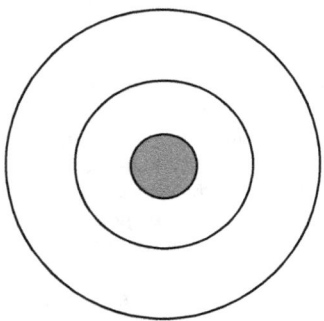

Figure 8-1

8.1 Particle Waves and Torque Distortions

波状螺旋网格形变被障碍物反弹便形成了粒子波。

When the waves of Torque Grids' distortion are bounced from a barrier, particle waves are formed:

$$mV = h/\lambda = h\upsilon$$

m：质量，V 速度，h：普朗克常数，λ：波长，v：频率

m: mass, V speed, h: Planck constant, λ: wavelength, v: frequency

8.2 Particle Waves and Torque Distortions

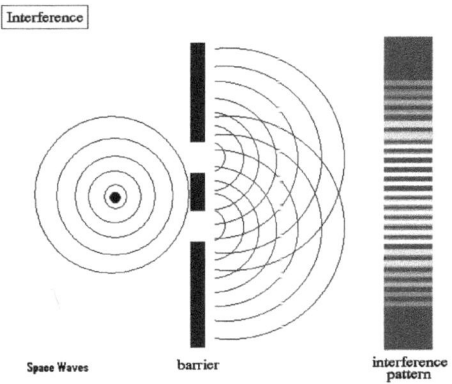

Figure 8-2

在以上双缝实验中，波状螺旋网格形变与两个狭缝作用产生了通过狭缝越过屏障的粒子波。

In the above Two Slits Experiment, the torque waves interact with the slits and generate particle waves beyond the barrier through two slits.

8.3 Wave–Particle Duality

在物理和化学中，波粒二象性概念是，所有的物质都具有波和粒子得特征。在量子力学中，粒子波被称为"概率波"。而统一场论则认为因波状螺旋网格形变而产生的粒子波是物质都具有波动性的原因，而并非因为粒子本身是波。在双缝实验中，当粒子通过狭缝后，其波状螺旋网格形变将与粒子波发生共振。而粒子波强的方向更容易发生共振。粒子向共振方向移动。

In physics and chemistry, wave–particle duality is the concept that all matter exhibits both wave-like and particle-like properties. In Quantum physics, the concept of "particle wave" or "probability waves" is used to study the wave-like properties of the particle. In this Unified Theory, the wave shaped Torque Grid distortion is the reason behind the wave-like properties of matter, regardless if the particle itself is wave or not. In two slits experiment, after passing through the slits, the particle's wave shaped Torque distortion will resonance with the particle waves. It is easier to have stronger resonance in the directions that have stronger particle wave. The particle moves along the direction of resonance.

9 Conclusion

以下是本理论的要点：

1. 本文提出的统一场论，给物理最基本的问题作出圆满的解释。而理论的精髓在于时间，空间和能量的概念密不可分的关系。而螺旋网格则把三种概念归纳为一体。

2. 能量使局部螺旋网格变大，当畸变与螺旋网格大小相同时，畸变与原来的螺旋网格谐振。螺旋网格实现共振的趋势产生基本的物理力，并可得到基本物理常数和定律。

3. 宇宙的苍穹是一个上层宇宙微小的螺旋网格。我们的宇宙中的最小单位是螺旋网格，而这一最小的螺旋网格是一个像我们的宇宙一样的宇宙。

4. 电子因在其壳上和螺旋网格上去畸变而稳定，而电子的共振是最基本的共振。因而，电子是一切速度低于光速的最基本粒子。比电子质量大的粒子都是以电子质量为单位的能量波的共振得以稳定而形成的。

5. 稳定和变化共存。能量产生螺旋网格畸变，从而带来不稳定性；而能量生成的螺旋网格畸变的共振带来了暂时的稳定。不存在永恒的稳定，也不存在永恒的不稳定。时间，空间和能量是永恒的。出生和死亡，创造和湮灭也永远不会结束。

The following highlights the theory:

1. The proposed unifies theory explains all physics theories. The essence of this theory is the inseparable relationship among time, space and energy. The Torque Grid unifies the three concepts.

2.   The energy enlarges the surrounding Grids. When the distortion is the same as the Torque Grid size, the distortion resonant with the original Torque Grid. Torque Grid's tendency of achieving resonance creates basic Physics forces and it derives basic Physics constants and laws.

3.   The firmament of the universe is a tinny universe Torque Grid in the vacuum of an upper grid hierarchy. The smallest unit in our universe is a Torque Grid which is a universe resembles to our universe.

4.   The electron has resonance points on its shell with Torque Grid, therefore, electron has the most basic resonance and it is the most basic particles moving at speed lower than the speed of light. The particles more massive than the electron are formed as result of the resonance of energy waves with the electron mass as their unit.

5.   Stabilization and change coexist. Energy is the source of distortion, hence the source of instability; resonance of energy introduced Torque distortion creates temporary stability. There is neither a permanent stability, nor a permanent instability. The space-time-energy is timeless. Therefore, birth and death, creation and annihilation never end.

10 Mathematical Analysis

10.1 Symmetry in Physics

在物理中有三种对称性：C, P 和 T

There are three symmetries in physics: C, P, and T.

Z 螺旋和 S 螺旋是 P 对称。或称镜像对称。

Z Torque and S Torque are P-symmetrical. They are mirror images of each other.

C 对称用于研究正负电。在本书中，P 对称代替了 C 对称。

C-symmetry and P-symmetry are the same concept in this book since C-symmetry is for charge and can be replaced by P-symmetry.

因为螺旋理论不支持黑洞理论。当然不承认为解释黑洞理论而发明的 T 对称。

Since the Torque theory does not support the black hole theory, it does not support theory of T-symmetry either.

因而，除了 P 对称外没有别的对称。也可以说，除了螺旋运动外没有别的三维运动。

Therefore, arbitrary 3D movements are P-symmetrical movements. In other words, Torque movements are the form of 3D movement.

10.2 Gravity Field

如果不考虑相互作用的影响，质量 M 产生的螺旋网格引力畸变为 x，质量 m 产生的引力畸变为 y，总的螺旋网格引力畸变为 x+y。如果考虑一个畸变是在另一畸变的基础上产生的，总的螺旋网格引力畸变为：

Disregarding interactions, mass M has Torque Grid gravity distortion x and mass m has Torque Grid gravity distortion x. The total Torque Grid gravity distortion is x+y. If the Torque Grid distortion x is on top of the second distortion y, the total distortion is:

$(1+x)(1+y) - 1 = x+y + xy$

因俩个畸变的相互作用效应而增加的畸变为：

The additional distortion from the interaction effect of Torque Grid distortion:

$(x+y + xy) - (x+y) = xy$

就是相互作用（引力势）能。

is the interaction (gravity potential) energy.

如果质量 m 到质量 M 的距离是 R，从质量 M 的总畸变为 aMR。因球面的面积为 $4\pi R^2$，而单位面积的畸变为 bM/R （b=a/4π）。

If the distance of the mass m from the mass M is R, the total distortion from M is aMR. Since the sphere's area is $4\pi R^2$, the distortion per unit area is bM/R (b=a/4π).

x= bM/R

质量 m 在一个螺旋网格的畸变为：

Mass m's distortion on a single Torque Grid is:

y=Km

E 就是引力能量：

E is the gravitational energy:

E=kxy=kKm(bM/R)= kKbmM/R

假设：

Assume:

G=kKb

即：

Then:

E=GmM/R

根据能量和力的关系，

Based on Newton's law of energy,

F=dE/dR= -GmM/R^2

以上负号表示力的方向同 R 增加的方向相反。

A negative value in the above equation indicates that the force is in the opposite direction of increment of R.

假设：

Assume:

f = -F,

$$f = \frac{GmM}{R^2}$$ (10-1)

以上是牛顿万有引力方程。

The above is Newton's gravity equation.

10.3 Torque Grid Size

当质量为 m 的粒子的波长为 D，而螺旋网格的大小也是 D 时，螺旋网格形变后的大小是 2D。由于螺旋网格畸变改变了空间时间和能量，实际质量减为 m/2。缺失的另一半变为势能。这是一个理论的例子，不会发生在现实中。

When the particle with mass m has wavelength of D and the Grid size is D, the enlarged Grid size is 2D. The wavelength is D. Since the Grid distortion changed space-time-energy, the actual mass is m/2. Half of the lost mass converted to gravitational energy. This is a theoretical case that never occurs in reality.

$GmM/(2D)=(1/2)mc^2$

$GmM/D=mc^2$

所以：

Therefore:

$D=GM/c^2$

普朗克方程(在这种特殊情况下，有圆形的波动路径)：

Planck equation (In this special case, the wave moves along a circular path):

$Mc^2=hc/\pi D$

即：

Or,

$M= h/(\pi Dc)$

$D=GM/c^2 =G(h/ (\pi Dc))/ c^2=Gh/(D\pi c^3)$

$$D = \sqrt{\frac{Gh}{\pi * c^3}} = 2.2856509 * 10^{-35} m \quad (10\text{-}2)$$

在弦论中的空间单位是普朗克长度 l_p 。以下是螺旋网格长度和普朗克长度之间的关系：

The unit of space in String Theory is Planck length l_p. Following is the relationship between the size of Torque Grid and Planck Length:

$$D = \sqrt{2}l_p \quad (10\text{-}3)$$

10.4 Electron Distortion

F_e 是在外壳上的扭力。n_0 是从中心螺旋网格到外壳间的螺旋网格数。r 是粒子的半径。

F_e is the force on the shell. n_0 is number of the Torque Grid between Central Grid and shell. r is the particle radius:

$$r = 2.81794 * 10^{-15} m \quad (10\text{-}4)$$
$$n_0 = 1.232883 * 10^{20} \quad (10\text{-}5)$$
$$F_e = 29.053510953 \text{N} \quad (10\text{-}6)$$

下面模型可用于计算电子在其外壳上的螺旋网格形变：

The following model can be used to calculate electron torque distortion on its outer shell:

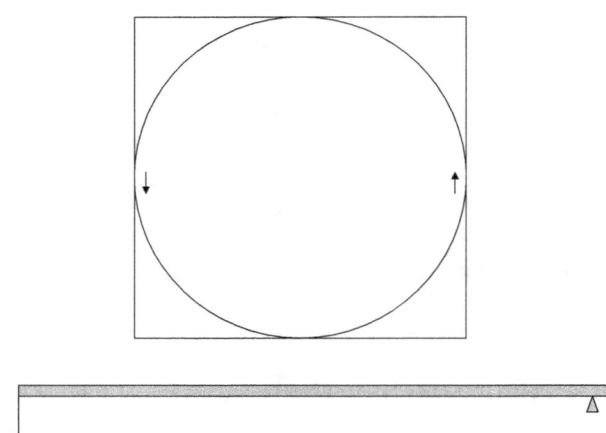

乘上 2π 把半径变成圆：

A factor of 2π is added to convert radius to circle.

$$d = 2\pi \frac{F_e}{F} \int_\pi^{n_0} (n_0 - n)\frac{n_0 - n}{n}Ddn \qquad (10\text{-}7)$$

$$d \approx 2\pi \frac{F_e}{F} n_0^2 D\left(\ln(n_0) - \ln \pi - \frac{3}{2} \right)$$

$$F = \frac{Gm^2}{D^2} = \frac{G}{D^2}\left(h / \pi Dc\right)^2 = \frac{Gh^2}{\pi^2 D^4 c^2} = 1.210339 * 10^{44}\,N$$

d=2π (29.053511/(1.210339*10^{44}))*(1.232883*10^{20})^2
(ln(1.232883*10^{20}) -lnπ − 1.5)

=0.999917767 D

$$d \approx D$$

在电子壳的总扭曲是螺旋网格的长度。

The total distortion on the electron's shell is proximately one Torque Grid size.

10.5 Electron Resonance

根据电子经典半径的定义：

Based on the definition of the electron's classic radius:

$$\frac{e * e}{r} = Mc^2$$

因为，

Since,

$$F_e = \frac{e * e}{r^2} \text{ and } r = n_0 D$$

所以，

Therefore,

$$F_e = \frac{Mc^2}{r} = \frac{Mc^2}{n_0 D}$$

$$1 \approx 2\pi \frac{Mc^2}{F} n_0 \left(\ln(n_0) - \ln \pi - \frac{3}{2} \right)$$

对每个 M 值，都可找到 n_0 以满足以上方程。还要加一个条件才能解出 M 和 n_0 。 这个条件就是电子半径与波长的关系。

For any value of M, there is a value n_0 to meet the previous condition. The additional relationship is needed to solve M and n_0. This condition is the relationship between the radius of the electron and its wavelength.

电子能量波长的一半是电子直径的 137.036π 倍，即电子周长的 137.036 倍，如同电子的园状波在电子轴上滚动。数值 0.036 是方形和圆形状的修正因子，类似于在前面的小节电子关于电子壳的总扭曲校正因子 $(1 - 0.999917767) * 137\pi = 0.035$ 。

The energy wavelength of the electron is 137.036π of electron's diameter, or 137.036 of electron's circumference, as if the electron' circular wave wheel roll on the diameter of the electron's axis. The value 0.036 is square/circle shape correction factor, similar to electron shape correction of $(1 - 0.999917767) * 137\pi = 0.035$ in the previous subsection for electron Torque distortion on its shell.

经过修正后的数字是 137。它是一个素数并有以下波动式：

After the correction, the number is changed to 137. It is a prime number and has following wave formula:

$3*5*8 + 3*5 + 2 = 137$

137 是存在波动式的最小素数。

The number 137 is the smallest prime number with proper wave formula.

例 如 ， $(2*3*5+2*3+2=38)$， $(2*2*4+2*2+2=22)$ 和 $(3*4*7+3*4+3=99)$ 不是素数 。

I.e., $(2*3*5+2*3+2=38)$, $(2*2*4+2*2+2=22)$ and $(3*4*7+3*4+3=99)$ are not prime number.

有了上述的电子半径与波长的关系，电子的 M 和 n_0 值是唯一的。

With the help of the above electron radius and its wavelength relationship, electron's M and n_0 values are unique.

10.6 Galaxy Energy Distribution

引力场是由螺旋网格的畸变所产生的。力乘距离为能量，而引力乘平均畸变为虚能量。

The gravity field is the result of energy distortion on the grid. Force times distance is energy. The gravity for unit force multiplies average distortion is virtual energy.

当波长为 L 的单位质量在半径为 r 的球体中，在球体上的平均畸变是球体的面积除以总畸变。

When the unit mass with wavelength L is in a sphere with radius r, the average distortion s on the sphere is the total distortion divided by the area of the sphere.

$ds=(D/L)dr/(4 \pi \ r^2)$

D: 螺旋网格长度

D: Torque Grid size

根据普朗克公式(单位质量)：

Based on the Planck equation, for unit mass:

$1*c^2=hc/L$

$ds=(Dc/h)dr/(4 \pi \ r^2)$

F 是两个有相质量 M 的物体的总虚拟引力：

F is the total virtual gravitational force between the two objects with the same mass M:

$$F = 4\pi r^2 \frac{GM^2}{r^2} = 4\pi GM^2$$

虚能量可定义为虚拟力乘畸变:

Virtual Energy is defined as the amount of virtual force multiplied by the distortion value:

dE = Fds = 4 π G M^2 *(Dc/h)dr/(4 π r^2)

$$dE = \frac{GDcM^2}{r^2 h} dr$$

当有相同质量 M 的两个星体拉开 dr 时,两星体的能量就增加了。每个星体得到能量增长的一半。当能量增长与虚能量变化相等时:

When two celestial objects with the same mass M are pulled part by dr, their energy increases. Each object shares half of the energy increment. When the energy increment equals to virtual energy change:

$$dMc^2 = \frac{GDcM^2}{2r^2 h} dr$$

$$d\left(\frac{1}{M}\right) = \frac{GD}{2ch} d\frac{1}{r}$$

定义常数 g 以简化上式:

Simplify the above equation by defining a new constant g,

$$g = \frac{2hc}{GD}$$

g=2.63912*10^{20} kg/m

微分方程的解是：

The solution of the differential equation is,

$$\frac{1}{M_0} - \frac{1}{M} = \frac{1}{g}\left(\frac{1}{r_0} - \frac{1}{r}\right)$$

当 M_0 和 r_0 相对较大，以上方程可简化为：

When M_0 and r_0 is relatively large, the above equation can be simplified to:

$$\frac{1}{M} = \frac{1}{gr}$$

Or:

M=gr

以上方程基于虚实能量相等的假定。由于螺旋网格的畸变代表了时空能的线性变化，因此，真实能量的分布以下述关系同步于虚能量：

The above equation is based on assumption that the virtual energy equals to actual energy. Since the Torque distortion represents the space-time-energy change, therefore, the actual energy allocation resonances with virtual energy as follow:

PM=Qgr

P 和 Q 是不为零的自然数。

P and Q are both non-zero positive integer.

假设 Q/P = K,

Assume Q/P = K,

$$M = Kgr \qquad (10-8)$$

10.7 Galaxy Rotational Curve

在旋臂星系里，旋转产生的离心力和引力是相等的：

In Spiral Arm galaxy, the rotational centrifugal force and gravity force are equal to:

$V^2 /r = GM/r^2$

$V^2 = GM/r = GKgr/r = KGg$

$V = (KGg)^{1/2} = K^{1/2}$ 131.5km/s

假设 $k = K^{1/2}$,

Assume that $k = K^{1/2}$,

$V = k*131.5km/s$

在一个旋臂的星系里，旋转速度是一个常数。

In a Spiral Arm galaxy, the rotation speed is a constant.

10.8 Black Hole and Thermal Dynamic Theory

引力场能捕获光子吗？ 引力场捕获光子的条件是引力 GMm/R^2 大于离心力 mC^2/R，即 $GMm/R^2 > mC^2/R$。可简化为：

Can a gravity field trap a photon? In order for gravity to tras a photon, the gravity force GMm/R^2 is greater then the reactive centrifugal force mC^2/R, or $GMm/R^2 > mC^2/R$. It can be simplified to:

$GMm/R > mC^2$

当质量 m 进入质量 M 的引力场的半径 R 处，逃逸能量 GMm/R 来自能量 mC^2。根据能量守恒定律，GMM/ R 不能大于原来的能量，或（GMM/ R）<MC2。因此，引力不能捕获光子；否则，会与能量守恒定律相矛盾。

When mass m enters to gravity field of mass M at radius R, the escape energy GMm/R is the originated from the energy mC^2. According to the law of energy conservation, the GMm/R cannot be greater than the original mass, or $(GMm/R) < mC^2$. Therefore, gravity cannot trap a photon; otherwise, it will contradict to the law of energy conservation.

虽然引力不能捕获个光子，但高密度的死恒星"黑洞"却提供了研究熵的一个极端情况。

Although the gravity cannot trap a photon, but the high density dead star "black hole" can provide an extreme case to study entropy.

一个粒子质量为 m 的走入质量为 M，在半径 R 处 $GM/(RC^2)$ 为 0.99 的"黑洞"。如果粒子到了半径 R 处而没有发生碰撞，它的质量的 99% 转化为动能。当粒子散失其速度，动能主要转化成光或一些黑物质，而粒子保留其原质量的 1%。其速度损失是由于热动力学第二定律："任何不处于热平衡状态的孤立系统的熵几乎总是增加。"但粒子只有其质量的 1%，而 99% 的质量是低熵的光子。这个结果显然违反了热动力学的第二定律。

A particle with mass m moves toward the "black hole" with mass M and at radius R where $GM/(RC^2)$ is 0.99. If the particle reaches radius R without collision, 99% of its mass is transformed to kinetic energy. When the particle losses its speed, the kinetic energy transformed mainly to light or possibly some Dark matters, while the particle keeps 1% of original mass. The lost speed is due to the second law of Thermal Dynamics: "The entropy of any isolated system not in thermal equilibrium almost always increases." But the particle only has 1% of its mass while 99% of the mass are photons with low entropy. This result clearly violates the second law of Thermal Dynamics.

首先，基于上述分析，可以得出这样的结论，热动力学第二定律只适用于一个粒子的质量没有被转换为其它能量形式的封闭系统。由于定律有了附加条件，其预测不再有效。虽可预测恒星会死，却不能预测宇宙会死。

First, based on the above analysis, it is concluded that the second law of Thermal Dynamics only applicable in a closed system where particle's mass is not transformed to other energy form. Since there is a condition attached to the law, its predictions are no longer valid. It can predict that a star will die one day, but is cannot predict that the universe will die one day.

其次，在上述情况下，死恒星"黑洞"毁灭了粒子 99%的能量，以一个更原始的形式，光子，释放能量。换句话说，能量再生了。恒星每天越来越老，但宇宙既不变年轻，也不变老。以螺旋网格扭曲的形式存在的能量从一个地方移动到另一个地方而不断转换，但宇宙却是永恒的。

Next, in the above case, the dead star "black hole" annihilates 99% of the mass of the particles and release the energy as photons in a more primitive form. In another words, the energy is re-born. The stars may continuously age, but universe doesn't, since the universe is neither getting younger nor older. The energy in the form of Torque Grids distortions movement from one place to the other and constantly transform, while the universe remains timeless.

10.9 Mechanism of Red-shift and Challenge to BBT

光子沿螺旋网格扭曲方向移动。因宇宙螺旋是右手松螺旋，光子也服从右手法则如下：

A photon moves forward along the Torque twisting direction. Since universal Torque is right handed loosening Torque, photon twisting movements follow right-hand rule.

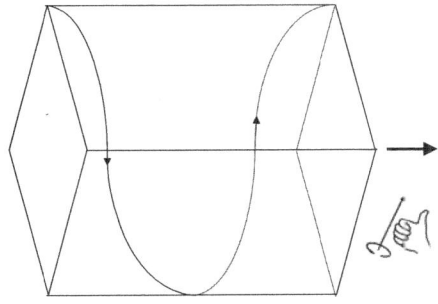

Figure 10-1

光子和单个电子都是对称的。在相互作用过程中光子的移动方向不变，因为只有电子沿光子的移动方向的动量与光子相互作用。因此，仅沿光子的移动速度方向的分量 Vy 的值增加了 Vc。

A photon and a single electron are symmetrical. The photon stays in the same moving direction during the interaction, because only electron momentum along the photon's moving direction Vy is interacting with the photon. Therefore, only the speed vector Vy along the movement of photon increases its value by Vc.

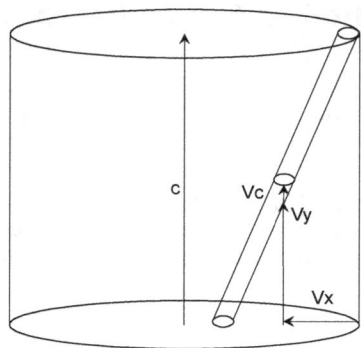

Figure 10-2 Interaction

假设 M 是一个电子的质量，m 为一个光子的质量互作过程中丢失或增加的部分，Vc 是在交互过程中的电子的速度变化，可得出下面的能量守恒，动量守恒方程：

Assume that M is the mass of an electron, m is the portion of a photon's mass lost or increased during the interaction, Vc is the speed change of electron during the interaction, then, there are following energy conservation and momentum conservation equations:

$$M(V_x^2 + (V_y+V_c)^2 - (V_x^2 + V_y^2))=mc^2$$

$$MV_c=mc$$

即，

Or,

$$V_c =c- 2V_y \qquad\qquad (10\text{-}9)$$
$$m = M(c - 2V_y) /c \qquad\qquad (10\text{-}10)$$

从方程式（10-10），

From formula (10-10),

$$V_y= 0.5c(1- m/M)$$

After the interaction, the new speed along the photon moving direction is greater than c/2:

Vc + Vy = c-Vy

对于可见光的红移，失去了质量小于原来的质量，m/M < $5.55 * 10^{-6}$，或：

For the visible light red-shift case, the lost mass is less than the original mass, m/M < $5.55 * 10^{-6}$, or:

$$0.5c(1 - 5.55 * 10^{-6}) < V_y < 0.5c$$

10.9.1 Collision Tube

对于固定的碰撞角，电子的轨道的电子/光子的碰撞过程中，形成的管装轨迹（图 10-2）。

For the fixed collision angle, during the electron/photon collision, the track of electron forms a tube (fig.10-2).

假设光子的波长λ，则碰撞目标的半径是λ，碰撞的可能性是：

Assume that the wavelength of the photon is λ, then the radius of the impact target is λ and the possibility of the collision is:

$A* \lambda^2$

由于碰撞管的横截面是固定的，相对于总的光子能量的光子的能量变化率是管横截面除以光子的横载面，或者：

Since the collision tube' cross section is fixed, the energy change ratio of the photon relative to the total photon energy is cross section of tube divided by cross section of photon, or:

B/ λ^2

平均能量的变化是：

The average energy change is:

AB=K

其中，K 是一个常数。

where K is a constant.

因此，不同光子的波长，相对于原来的光子能量的平均变化率是相同的。

Therefore, regardless of the photon's wavelength, the average change percentage relative to the original photon energy is the same.

10.9.2 Crossing Speed

图 10-2 中的 Vx 的被称为跨越速度。它决定的光子的碰撞管的角度。假设 VY= C/2，光子的波长 is λ，碰撞持续时间为 t，

Vx in Fig 10-2 is called crossing speed. It decides the angle of the impacted tube of the photon. Assume Vy=c/2, the wavelength of photon is λ, collision duration is t,

$\lambda + (1/2)ct = ct$

$$t = 2\lambda/c \qquad (10\text{-}11)$$

电子与光子相互作用持续时间是光子周期λ/c 的两倍。当 Vx> C / 2 时，电子可以穿过光子从一侧到达光子的另一侧。在这种情况下，高速电子迫使碰撞提前结束。电子显得比光子"更快"。总的门槛值速度是：

The electron/photon interaction can take twice the photon's period λ/c. When Vx > c/2, the electron can cross from one side of photon to the other side of the photon. In this case, the high speed electron forces the completion of collision ahead of normal "schedule". Therefore, the electron seems to be "faster" than the photon. The total threshold speed is:

$$V = \sqrt{(c/2)^2 + (c/2)^2} = \frac{\sqrt{2}c}{2} = 0.7071c \qquad (10\text{-}12)$$

当电子的速度大于 0.7071c，电子开始释放能量的光子而光子波长开始蓝移"赶上"电子。

When the speed of electron is greater than 0.7071c, the electron starts to release energy to photon and photon wavelength starts to blue-shift to "catch up" with the electron.

10.9.3 Blue-Shift

当 V>0.7071c 时，光子的质量增量遵循（10-10）。

When V>0.7071c, the photon mass increment follows (10-10).

在这种情况下，质量增量 m 的可能有更大的未知范围。

In this case, the increased mass m may have a bigger unknown range.

10.9.4 Re-Evaluate Red-Shifting

光子与电子的相互作用是天体红移的来源之一。如果多普勒效应不是红移的唯一的原因，那么宇宙大爆炸理论就失去了有力的证据。

The photon electron interaction is one of the sources of red-shifting of celestial object. There is no strong evidence for Big Bang theory if Doppler effects are not the only reason for red-shifting.

10.9.5 Re-Evaluate the Other BBT Bases

10.9.5.1 Remote Galaxies are Not Far Apart

宇宙大爆炸的另一个重要根据，是遥远的星系相距甚远。

Another important base for Big Bang is that the remote galaxies are far apart.

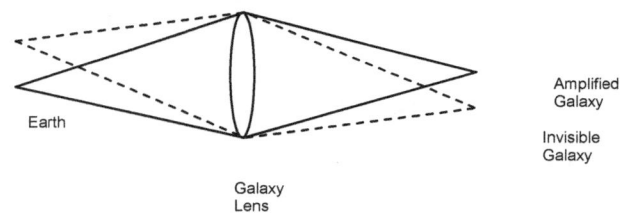

引力透镜效应，使一些星系清晰了，但它却使许多星系不见了。结果，由于光学的错觉，可见遥远的星系相距甚远。

The gravity lensing makes some galaxies clearer, but it makes many Galaxies invisible. As the result, the visible remote galaxies are far apart due to the optical illusion.

10.9.5.2 No Evolution at Universe Level

BBT 理论使用星系演化做其证明。螺旋星系没有不规则星系和椭圆星系那么明亮。如果星系太远，那么，螺旋星系就不太容易看到。这种光学幻觉，使我们相信，遥远星系不及俯近的星系成熟。由于望远镜变得越来越好，越来越遥远星系将被认定为螺旋星系。

BBT theory uses galaxy evolution as its proof. The Spiral Galaxy is less bright than irregular galaxies and elliptical galaxies. If the galaxies are too far, then, the Spirals are less visible. This impression makes us believe that the distant galaxies are less mature than the closer galaxies.

As the telescope becomes better and better, more and more distant galaxies will be identified as spiral galaxies.

10.9.5.3　Las Campanas Red-Shift Survey

Las Campanas 的红移勘测采用光子的红移为手段，以测量距离，移动速度和星系的分布。它得出结论认为，宇宙有大规模的同一性。

Las Campanas Redshift Survey uses photon's red-shift as means to measure the distance, moving speeds and distribution of the galaxies. It concludes that the universe has large-scale homogeneity.

电子光子相互作用造成了比多普勒效应更大的红移。遥远星系的速度不能被准确测量。此外，由于引力的透镜效应，从观察的角度，遥远的星系被扩大或偏转消失。大规模的同一化是从地球的角度看到的一个万花筒对称错觉。

The electron photon interaction causes a bigger red-shift than Doppler effects. The travelling speed of the remote galaxies can no longer be accurately measured. Due to the gravity lensing, the remote galaxies are enlarged or deflected out of the view based on the angle of observation. The large-scale homogeneity is a symmetrical Kaleidoscope optical illusion from earth stand point of the view.

10.9.5.4　Hubble Diagram

哈勃图用红移为依据给出距离和速度之间的关系。由于遥远星系的速度不能被准确测量，这样的关系没有任何意义。

Hubble Diagram uses red-shift as means to provide the relationship between the distance and velocity. Since the velocity of the remote galaxies can not be measured accurately, such relation has no meaning.

10.9.6 Conclusion

统一场论挪去所有 BBT 依据，使 BBT 不再可信。

Unified Field Theory put BBT in doubt by removing all the bases from BBT.

10.10 The Limit of Gravity and the Universe Grid

基于前一章 GM/RC^2 〈1，如果宇宙有一个恒定的密度 d (9.22*10^{-27})，那么，

Based on the previous chapter, GM/RC^2 〈1, if the universe has a constant density d (9.22*10^{-27}), then,

$$GM/(RC^2) = (3/4 \; \pi) \; R^3 \; GH/(RC^2) = (3/4 \; \pi) \; R^2 \; Gd/C^2$$

当它等于 1 时：

When it equals to one:

$$(3/4 \; \pi) \; R^2 \; Gd/C^2 = 1$$

$$R = (C^2 /((3/4 \; \pi) \; Gd))^{1/2} \; =2.49 * 10^{26} \; m$$

如果宇宙确实有一个恒定的密度，当宇宙达到 2.49 * 10^{26} m 时，宇宙达到一个极限。此极限有几个含义。首先，如果空间是有限的，那么，它假设地球是宇宙的中心在这个距离达到的极限。其次，如果宇宙无限，那么，我们的宇宙存在依赖于其他宇宙。

When the universe reaches to 2.49 * 10^{26} m, the universe reaches to a limit, if the universe does have a constant density. This limit has a few meanings. First, if the space is limited, then, it reaches to the limit at this distance assuming the earth is the center of universe. Second, if the universe is not limited, then, the reason for the existence of our universe relies on the other universes.

根据目前的数据，我们可以看到我们的宇宙极限外的星系。因此，唯一可能的模式是，宇宙是无限的。

Current data support that we can see galaxies outside of our universe limit. Therefore, the only possible model is that the universe is infinite.

$2.49 * 10^{26}$ m 是宇宙螺旋网格的大小，这是一个在较高层次的螺旋网格。每个网格在它自己的宇宙中有相同的能量密度，以达到稳定。

The value $2.49 * 10^{26}$ m is the logical size of the universal Grid which is a Grid in the higher Torque Grid hierarchy. Each Grid has same energy density in its own universe to achieve stabilities.

宇宙有没有明显的边界。网格的边与内部将没有任何区别。

The universe has no visible boundary. There will be no difference between being of the border and being inside the Grid.

10.10.1   Energy-Time-Space

宇宙中的能量密度决定宇宙的网格大小。网格的大小与光速决定的空间和时间。因此，能源，空间和时间是分不开的。

The energy density in the universe decides the universe's Grid size. Grid size plus light speed defines space and time. Therefore, energy, space and time are inseparable.

10.10.2   Hierarchy Ratio Constant

螺旋网格尺寸 D 乘 N 是宇宙的网格大小。宇宙网格大小乘 N 更上一级的宇宙的网格大小。同样，螺旋网格大小除以 N 是下一个级别螺旋网格的大小。根据今天所知的宇宙的密度：

The Torque Grid size D times N is the universe Grid size. The universe Grid size times N is the Grid size of the next level Grid above the universe. For the same reason, the Torque Grid size divided by N is the size of Torque Grid one level lower than the Torque Grid. Based on the density of universe known today:

$N=(2.49 * 10^{26})/(2.2856509 * 10^{-35})= 1.0894* 10^{60}$

10.10.3   Universal Grid in the Vacuum

我们的宇宙很可能在较高的螺旋网格的层次结构的在真空里。

Our universe can be a Grid in the vacuum at the higher Torque Grid hierarchy.

宇宙的密度是 $9.22*10^{-27}$。在较高的网格层次结构，其能量密度是相同的。如果随机挑选一个网格，不是在真空中的可能性是 $9.22*10^{-27}$。这一推论使这纯粹是基于统计预测。当然，能源对扭矩网格的影响非常小。

The density of the universe is $9.22*10^{-27}$. In the higher Grid hierarchy, the energy density is the same. If a Grid is randomly picked, the possibility of not being in the vacuum is $9.22*10^{-27}$. This prediction is purely based on statistics. Never the less, energy has very little impacts on Torque Grids.

10.11 Torque Waves

让我们把话题从宇宙转到粒子。

Lets switch our topic from the universe to the particle.

由于电子具有最小质量，其它粒子内部的基本波 W1 有比电子较大的质量。

A basic wave W_1 inside other particles has a larger mass than that of the electron, since the electron has the smallest mass.

电子有单位电荷。粒子的基本波 W_1 以电子为基础，须与电荷扭曲谐振。因此，它带一个单位电荷。

The electron has unit charge. The particle's basic wave W_1 is based on the electron and needs to resonate with the charged distortion. Therefore, it contains a unit charge.

单位电荷能量也在电子壳上产生 1/137 螺旋网格大小的畸变。W1 也在壳电子产生畸变。为了与在电子壳上单位电荷能量的畸变谐振，因此，质量 W1 为正整数 S1 乘以电子的质量，畸变为 S1/137。由于 137 是一个素数，它使共振简化。

Unit charge energy creates a distortion with 1/137th the size of the Torque Grid on the electron shell. W_1 creates distortion on shell of electron. In order to be in resonance with unit charge distortion hence, the mass of W_1 are integer number S_1 multiplies electron mass to make distortion to be $S_1/137$. Since 137 is a prime number, it simplifies the resonance.

10.12 Torque Wave Interaction

螺旋波可用一般的数学公功式表达为：

Torque waves can be expressed mathematically in generic forms as in the following functional expression:

$E_n(S_n, E_{n-1} (S_{n-1}, \ldots E_1(S_1, S))\ldots))$

强相互作用表达为下列乘法：

The strong interaction is expressed in a geometric series as follow:

$E_n (S_n , E_n (\ldots)) = S_n S_{n-1} \ldots S_1 S$

一旦 W1 形成，下个强相互作用波 W2 具有与 W1 不同的电荷，W1 的畸变 S1/137 是其基础。从新波 W2 看来，它有整数 S2 个 W1。总畸变为（S1* S2）/ 137。总畸变倍增。 N

个波总畸变为（S1* S2*...* SN）/ 137。为了保持谐振，
S1，S2，...，Sn 的值为整数，总电荷为-1，0 或 1。

Once W_1 forms, the next strong interacting charged wave
W_2 has different charge from W_1 and the distortion $S_1/137$ is
its base. From the new wave stands point of the view, it
has integer number S_2 of W1. The total distortion is
$(S_1*S_2)/137$. The total distortion multiplies. The n waves
total distortion can be $(S_1* S_2* ...*S_n)/137$. To maintain
resonance, the values of S_1, S_2 , ...,S_n, are integers and the
total charge is -1, 0 or 1.

非强相互作用的函数是下列的和：

The non-strong interaction function is the following
arithmetic series:

$E_n (S_n , E_n (...)) = S_n + S_{n-1} + ... + S_1 + S$

10.13 Strong Waves Interactions

普朗克公式：

Planck equation,

E=hv

使用普朗克常数作为能量单位，普朗克方程式变为：

Use Planck constant as energy unit, the Planck equation
becomes:

E=v

即，

Or,

$S_n = v_n$

螺旋波可用复函数表达为：

Torque distortion wave can be expressed as complex function:

$$S_n(x_n, t) = v_n e^{2\pi i (v\frac{x_n}{c} - v_t)}$$

两波强相互作用：

Two-wave strong interaction:

$$S_1(x_1, t)S_2(x_2, t) = v_1 v_2 e^{2\pi i(\frac{v_1}{c}.x_1 + \frac{v_2}{c}x_2 - (v_1 + v_2)t)}$$

多波的强相互作用：

Multiple waves' strong interaction:

$$S_1(x_1, t)S_2(x_2, t)... = v_1 v_2 .. e^{2\pi i(\frac{v_1}{c}.x_1 + \frac{v_2}{c}x_2 + ... - (v_1 + v_2 + ...)t)}$$ (10-12)

Use the one as the value of x_1 to x_n in formula (10-12):

$$S_1(1, t)S_2(1, t)S_3(1, t) = v_1 v_2 v_3 e^{2\pi i(v_1 + v_2 + v_3)(\frac{1}{c} - t)}$$

在三个波的情况下，如果第三频率包含前两波的频率：

In three waves' case, if the third frequency contains the previous two waves' frequency:

$$v_1 + v_2 = v_3$$ (10-13)

这样，三个波就稳定了。其原因是：

Then, the three waves are stable. The reason is that:

$$S_1(1, t)S_2(1, t) = v_1 v_2 e^{2\pi i(v_1 + v_2)(\frac{1}{c} - t)} = v_1 v_2 e^{2\pi i v_3 (\frac{1}{c} - t)}$$

And:

$$S_3(1,t) = v_3 e^{2\pi i v_3(\frac{1}{c}-t)}$$

有相同的频率。前两波相互作用产生的波与第三波共振。

Have same frequency: v_3. The result of the first two waves' interaction is in resonance with the third wave.

如果电子的能量为单位质量，总质量可以表示如下：

If the electron energy is the unit, the total mass can be expressed as follow:

$$E_{total} = v_1 v_2 ... v_n$$

10.14 Fibonacci Series

等式（10-13）的一般形式是：

General form of equation (10-13) is:

$$v_{n-2} + v_{n-1} = v_n \qquad (10\text{-}14)$$

电子的频率为单位。最小的螺旋波不能是一因为一乘其他值不会改变结果。因此，数列的起始值是二。数列的第二个值可以是二或三。当数列的第二个值是三时，数列是斐波那契数列：

The frequency of the electron is the base unit used to measure all other particle waves. The smallest Torque wave cannot be one, because it can't interact with other waves. Therefore, the series start with two. The second number in the series can be two or three. If the second number is three, the series are following Fibonacci series:

2, 3, 5, 8, 13, 21, 34, 58, 89, 144 …

10.15 Strong Forces

电子的波长是 2.42631021×10⁻¹² m。它是电子半径的 137.036π倍。这意味着，电子的质量在电子外壳上产生的沿轴线畸变是 1/137.036π网格大小。沿轴线的畸变转化为沿着电子的壳的畸变后增加了π倍，大小为 ˊ/137.036 网格长度。数值 0.036 方/圆形状的校正因子。调整后的畸变是网格大小的 1/137。

The electron's wavelength is 2.42631021×10⁻¹² m. It is 137.036π times greater than the electron's radius. The electron mass distortion along the axis on the shell of electron is 1/137.036π of the Grid size. After transforming the distortion from axis direction to along the shell, the distortion increases by factor of π. In order to fit the cylindrical wave into a spherical, the square/circular shape correction factor must be used. This is because the ends of the cylindrical wave are flat. The value 0.036 is a square/circular shape correction factor. The new adjusted distortion is 1/137 of Grid size.

强相互作用与电子外壳上产生的畸变密切相关。当一个强相互作用的波动数列的总质量大于或等于电子的质量的 137 倍，这 137 倍的电子质量的部分形成一个稳定的能量波。这种能量波在电子外壳上产生一个网格大小的畸变。它是在单位电荷畸变共振而变得稳定。要想拆散粒子，需要 137 多个电子的能量打破这个障碍。强力不是一种力，而是一种障碍能量。

The Strong force is closely related to electron's distortion on its shell. When the total mass of a strong interactive wave series is greater than or equal to 137 times the mass of an electron, the strong interactive wave forms a stable energy wave. This energy wave distorts the Torque Grid on the shell of electron, same as the unit charge. It is in resonance with the unit charge distortion and becomes stable wave by itself. To break up the particle, more than 137 times the electron's energy is needed to break up this barrier. The strong bonding force is not a force. It is an energy barrier.

10.16 Prime Number 137

数字 137 是一个素数，并具有以下的质量式：

The number 137 is a prime number and has the following mass formula:

3*5*8 + 3*5 + 2

斐波那契数列为：

With Fibonacci series:

2, 3, 5, 8

在可能的质量公式的数字中，137 号是最小的素数。

The number 137 is the smallest prime number among the numbers with a possible mass formula.

i.e.:

(2*3*5+2*3+2=38)

(2*2*4+2*2+2=22)

(3*4*7+3*4+3=99)

不是素数。

are not prime numbers.

10.17 Weak Interaction

素数 137 在电子的壳上把单位电荷分成 137 个小单位，它也将电子半径为 137 段。在二维圆形平面，它产生了 137 *137 个波。每个波有（1/137）*（1/137）的单位电荷能量。在中心的波与电子共享同一圆心形成一个球体。每个波的长度是 1/137 电子半径。

The gravity distortion prime number of 137 is used to divide the unit charge into 137 smaller units on the electron's shell, and it also divides the electron radius into 137 sections. In the 2D circular plane, it creates 137*137 waves. Each wave has energy of (1/137)*(1/137) unit charge. Each wave's length is 1/137 electron radius.

当两个波发生弱相互作用，因 137 是素数，仅有 137*137 波中的一个单元参与。其他的数字，如 2，3，4，不能整除 137 *137 与电子波产生共振。

When two waves weakly interact, only one wave energy unit out of 137*137 waves can participate the interaction since 137 is a prime number. The other numbers, such as 2, 3, and 4, can not divide 137*137 and resonate with the electron wave.

两个粒子之间的弱力是没有意义。与此相反，能量势垒更有意义并据可测试性。弱相互作用能为：

The weak force between two particles is not meaningful. On the contrary, the energy barrier can be more meaningful and tenable. Weak interaction energy is:

$(1/137)*(1/137) = 5.3 * 10^{-5}$

当一个粒子具有许多相互作用波时，每个带电波结构有一个波球。波球变成带电波结构的共振中心，没有两波球是可以重叠的。

When a particle has many interactive charged waves, each charged wave structure has a wave ball at the center. The wave ball becomes the resonance center of the charged wave structure and no two wave balls can be overlapped.

弱相互作用不会引起衰变。粒子衰变的根源是不和谐引起的波衰减，共振波结构的优胜过程。而能垒会减少启动衰变的可能性。

The weak interaction does not cause decay. The main cause of particle decay is dissonance caused by wave decay and resonance wave structure winning process. The energy barrier reduces the possibility of starting the decay.

与现有的标准模型理论相反，不能证明弱力在距离更近时会变得更大。根据波的性质，波强度朝在靠近向球心时不会改变。

Contrary to the existing Standard Model Theory, there is no proof that the weak force can get any larger when the range becomes closer. The wave strength in the center sphere does not change toward the center of the sphere based on the nature of the wave.

10.18 Energy Distribution in Particles

根据公式（10-7），单位电荷的螺旋能量可由下式计算（n 以螺旋网格大小为单位）：

According to formula (10-7), the unit charge torque energy can be calculated with the following (The n the number of Torque Grids):

$$torque = \frac{2\pi F_e}{F} \int_{\pi}^{n} \frac{n_0 - n}{n} dn = \frac{F_e}{F} n_0 (\ln(n) - \frac{n}{n_0})$$

$$\approx \frac{2\pi F_e}{F} n_0 \ln(n)$$

畸变能量，

And distortion energy,

$$distortion \approx 2\pi F_e / F \int_\pi^{n_0} ((n_0 - n)^2 / n) D \, dn$$

$$= 2\pi (F_e / F) n_0^2 D \big(\ln(n_0) - \ln \pi - 3/2 \big)$$

螺旋量与畸变能相比之下要小很多。能量分布仅决定于畸变能。

The torque energy is small compared to the distortion energy. The energy distribution is based on distortion energy only.

累积分布函数：

The cumulative distribution function:

$$\frac{n_0^2 \ln(n)}{n_0^2 \ln(n_0)} = \frac{\ln(n)}{\ln(n_0)}$$

$$\frac{\ln(n)}{47.5} \tag{10-15}$$

分布函数：

Distribute function:

$$\frac{n_0^2 / n - 2n_0 + n}{\ln(n_0)n_0^2} = \frac{1/n - 2/n_0 + n / n_0^2}{\ln(n_0)} \approx \frac{1}{n\ln(n_0)}$$

以下是电子能量分布函数（n 以螺旋网格大小为单位）：

The following is the electron energy distribution function (The n the number of Torque Grids):

$$\frac{1}{47.5n} \tag{10-16}$$

71

带电波具有电子的能量分布相同的能量分布而相互共振。

A charged wave has the same energy distribution as the electron energy distribution to be in resonance with the unit charge.

10.19 The Meanings of the Particle Formula

粒子的质量式是一个结构与成分配置。粒子的结构与成分配置不仅可用于解释已知的粒子的特性，而且可预测未知的颗粒的存在与特性。

A particle mass formula is a structural composition formula. The structural composition in a particle is useful not only in explaining the characteristics of the known particles, but also in predicting the existence and characteristics of the unknown particles.

2*2* 3，A =2* 3 * 5，B = 2* 2 * 4，A^2 和 B^2 是粒子基本成分的质量。粒子的质量公式是基本成分质量的总和。有些质量式可能有额外的成分，如在 T+ 中的 9* 10 *19。

2*2, 2*3, A = 2*3*5, B = 2*2*4, A^2 and B^2 are the masses of basic particle structural components. The mass formula for a particle is summation of the basic component masses. Some of the formula may be has additional components, such as 9*10*19 in Tauon.

含 A^2 和 B^2 成分的粒子是强子。其原因是强力的最小能量值为 137 并且不能被电荷约束。A^2 和 B^2 满足这一的要求。T+ 中的 9* 10 *19 成分部分满足这一要求，但其能量被电荷约束，因 9*10*19 的总电荷不为零。

The particle with A^2 and B^2 components are Hadrons. The reason is that the minimum strong force energy is 137 and it can not be bonded by charge. A^2 and B^2 meet such requirement. Tauon's component 9*10*19 partially meets such requirement, but the energy is not free, since 9*10*19 has a charge.

自反粒子的成分十分有趣:

Self Anti Particles formula are interesting:

η^0

Eta meson^0

$4*B^2 + 2*B + 8 + 8$

η'^0

Eta prime meson^0

$2*A^2 + 2*A + 7 + 7$

φ^0

Phi meson^0

$2*A^2 + 6*A + 6*2.5$

自反粒子的共同特征是，公式中的每一种类型的成分都有偶数个。

The common characteristic of self anti particle are that each type of the component in the formula is in even number.

10.20 Topology of Proton

电子是最稳定的粒子。第二稳定的粒子是质子。质子的结构是 2 * A2 + A +2*（30.076335）。可简化为 2 * A2+ A2* 3。

The electron is the most stable particle. The second most stable particle is the proton. The structure of proton is $2*A^2 + A + 2*(3.076335)$. It can be simplified to $2*A^2 + A + 2*3$.

73

10.20.1   Resonance and Shape

质子内部的波以多种方式共振，A ←→ A², A² ←→ A², 2*3 ←→A 等。

The internal Proton waves are in resonance in many ways, A ←→ A², A² ←→ A², 2*3 ←→A resonance, etc.

质子的主要结构成分是 A²，A² 和 A。主体结构中的每个部件沿着三维轴放置，x → A²， y → A² 和 z → A。三维轴与单位电荷壳轴有六个交点。连接外壳上的六个交点，就形成了对称的八面体结构：

The proton's main structural components are A², A² and A. Each component in the main structure is placed along a 3D coordinator axis, x → A², y → A² and z → A. The axes intercept with the unit charge shell with six total intercept points. Connect the six intercepting points on the shell will form the following symmetrical octahedron structure:

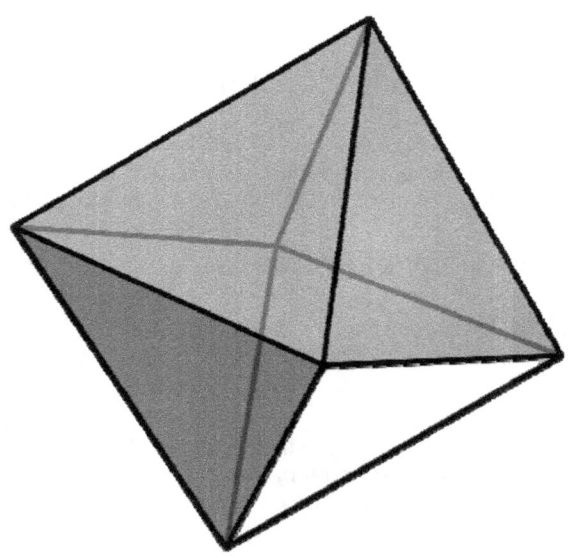

上述结构的质子尺寸是电子的大小近 50％。它不再是形球的。而是八个面。

The above structure makes proton size almost 50% of electron size. It is no longer ball shaped. It has eight faces instead.

10.20.2   Single Proton Charged Field

螺旋电力线比较容易通过八面体的八个面，因为那儿没有结构波，干扰少于三个轴。质子唯一的正电荷通过八个面，随机，均匀地与外部带电粒子相互作用。

It is easier for Torque lines go through the eight faces on octahedron since there are structural waves and give less interference on Torque as compared to three axes. The single positive charge Torque of Proton goes through eight faces randomly and evenly to interact with the external charged particles.

10.20.3   From Octahedron to Octal-Spots

当核的质子数小于三时，质子自旋与电子，通过八个面的正电荷电力线随机分布。这就可以解释为什么两个电子能使氢和氦稳定。

When the number of protons in a nuclear is less than three, the proton spins and the electrons can be randomly allocated. This can explain why two electrons make Hydrogen and Helium stable.

当有三个以上的质子，在原子中的质子不自旋。假设原子是一个虚拟的八面体。八面体的质子的每个面与原子八面体的一个面总是面向相同的方向。因此，电力线总是经过原子的八面。八个面的方向，可称为八点。要与八点共振，原子外层的轨道需要八个电子才稳定。

When there are more than three protons, the protons in the atom do not self-spin. The atom can be considered as a virtual octahedron. Each face of the octahedron proton always faces same direction of a face of atomic octahedron. Therefore, charged Torque lines always go through eight faces of atomic octahedron. The directions of eight faces are called Octal-Spots. To be in resonance with Octal-Spots, the outer orbit of the atom needs eight electrons to be stable.

由于其简单而灵活，s 和 p 电子云可以很容易地填补八点。而 d 和 f 电子云用他们的子云填八点。

The electronic cloud s and p can simply fill-in Octal-Spots due to their simplicity and flexibility. The cloud d and f can fill-in Octal-Spots with their sub-clouds.

10.20.4   Element Iridium Analysis

重金属外壳的 d 电子云不大僵硬，可以填充八点中的一些点。

In the heavy metal, the outer shell's d clouds are less rigid and can fill-in a few of Octal-Spots.

铱电子云的配置是 $1s^2$ $2s^2p^6$ $3s^2p^6d^{10}$ $4s^2p^6d^{10}f^{14}$ $5s^2p^6d^7$ $6s^2$。

The electron configuration for Iridium is $1s^2$ $2s^2p^6$ $3s^2p^6d^{10}$ $4s^2p^6d^{10}f^{14}$ $5s^2p^6d^7$ $6s^2$.

外层电子云有 $5d^7$ 和 $6s^2$。

The outer shell has $5d^7$ and $6s^2$.

$5d^7$ 电子云可以由 $2*dz^2$, $2*dxz$, $2*dyz$ 和 dxy 子云组成。电子云总数是 2+2 +5*4=24。 d 电子云平均有四个子云。电子云总数除四是合乎逻辑的电子数：24/4=6。$5d^7$ 电子云有六个合乎逻辑的电子数填充八点中的六点。

The 5d^7 cloud can be composed of 2*dz^2, 2*dxz, 2*dyz and dxy. The total count sub-cloud is 2+2+5*4 = 24. The average d cloud has four sub-clouds. The total count sub-cloud divided by four is logical electron count: 24/4=6. The 5d^7 cloud has six logical electrons which fill-in six of Octal-Spots.

6s^{2}电子云（第六层两个电子）填充八点中的两点。

6s^2 (shell 6 with 2 s clouds) fills-in the two of Octal-Spots.

现在，所有八点被占了。这使铱成为最具抗腐蚀性的金属。

Now, all Octal-Spots are occupied. This makes Iridium the most corrosion resist metal.

10.21 Proton Pile in Nuclei

10.21.1   Base Square

当质子数是四时，质子将形成 2×2 方形结构。虽然核不旋转，但是在核内的电场矢量不停旋转。在 2×2 的结构中，每个质子在波旋转过程中地位相同。在 3×3 的结构中，位与中心的质子的拓扑地位不同于其他质子。因而 3×3 基础方型结构不存在。

When the proton count is four, the protons will form a 2*2 square structure. Even though the nucleus does not rotate, the charged vectors in the nuclear are rotating. In a 2*2 structure, each proton is equally positioned in the wave rotating process. On the other hand, 3*3 does not work the same way. In a 3*3 structure, the proton in the middle is not topologically equal to the other protons. Therefore, 3*3 base square structure does not exist.

下一个可能的正方形是 4×4。这是一个每个元素为一个 2×2 正方形的 2×2 正方形。接下来是 8 × 8, 16 ×16, 32 × 32, 等等。在一般情况下，其公式为：

The next possible square is 4*4. It is a 2*2 square in which each element is a 2*2 square. Next is 8*8, then 16*16, 32*32, etc. In general, the formula is:

$2^n * 2^n$

10.21.2   Nuclei Stabilities

原子核中的波运动主要由基础方形结构控制。在其它层中的质子波运动从束于基础方形结构。质子对称地堆积着。但不是所有的核都可以是对称的。质子堆的对称性决定了原子核的稳定性。

The wave movements in nuclei are dominated by the base square. The wave movements in the other layers of the proton pile follow the base square. The protons are piled symmetrically. But not all the nuclei can be symmetrical. The symmetrical of the proton pile decides the stabilities of nuclei.

当某一元素的核因缺少一个质子未能形成完美的对称结构，其稳定同位素的数量较少。有时，它没有稳定同位素。

When an element's nuclear is missing one proton from forming perfect symmetrical structure, its stable isotopes' count is less. Sometime, it has no stable isotope.

锝有 43 个质子。质子堆的结构是：

Technetium has 43 protons. The proton pile structure is:

4*4 + 2*(3*3) + 2*(2*2) + 1

4 × 4 + 2 × （3 × 3）+2 × （2 × 2）是一个对称的结构，多出的质子使结构变成了非对称。因此，锝没有稳定的同位素。衰变后，要么从锝较轻的同位素变成钼：

4*4 + 2*(3*3) + 2*(2*2) is a symmetrical structure, additional Proton makes the structure non-symmetrical. Therefore,

78

Technetium has no stable isotope. After decaying, it will either be Mo from lighter Technetium isotopes:

4*4 + 2*(3*3) + 2*(2*2)

要么从锝较重的同位素变成钌：

Or, Ruthenium from heavier Technetium isotopes:

4*4 + 2*(3*3) + 2*(2*2) +2

另一个有趣的不稳定的元素是钷。质子堆的结构是：

Another interesting non-stable element is Promethium. The Proton pile is:

2*(4*4) + 2*(3*3) + 2*(2*2) + 2*1 +1

如果钷获得一个的质子或减少一个质子，核结构变得对称稳定。

If Promethium gains one more proton or reduces one proton, the nuclear structure becomes symmetrical and stable.

元素镨有以下结构：

The element Praseodymium has the following structure:

2*(4*4) + 2*(3*3) + 2*(2*2) +1

结构非对称，镨只有一个稳定的同位素。

The structure is non-symmetrical and Praseodymium only has one stable isotope.

10.22 Torque and Relativity

10.22.1　　Michelson–Morley Experiment

Michelson–Morley 实验是很多为相对论提供实验基础的理论子一。但正如穆罕默德·沙菲克汗的文章所表明，它是一个被误解和曲解的实验。

The Michelson–Morley experiment is one of many experiments that provide the experimental foundation to the theory of relativity. But, it is a misconceived & misinterpreted experiment as indicated in Mohammad Shafiq Khan's article.

Michelson–Morley 实验用的波长变化来判断行驶距离的变化。实验假定光通过以太时，以太风造成额外行驶距离，而距离的变化会导致波长的变化。

The Michelson–Morley experiment uses change of the wavelength to tell the change of travelling distance. It assumes that the aether wind caused additional travelling distance when light passes through aether and the distance change can cause the wavelength change as well.

我们可以用声音代替激光有，用空气代替以太，重复相同的实验。

We can replace laser with sound and replace aether with air to repeat the same experiment.

当两车以相同的速度行驶时，风可以使声音的两辆车之间传播时的路径不相同。第一次测试，A 汽车在前 B 车在后面，而两者都是在每小时 65 英里沿相同的高速车道行驶。声音在空气中行驶的距离从 A 车到 B 车较短。 A 车按响喇叭，从汽车 B 听到了声音。接下来，B 汽车在前 A 车在后面。A 喇叭车再次按响喇叭，从汽车 B 听到了声音。 B 车会发现，即使声音行驶距离改变了，从汽车 A 喇叭的声音的频率不改变。

When two cars are driving at the same speed, the wind can make sound path the between two cars different. First test, car A are at the front and car B at the back, while both are driving at 65 miles per hour along the same high way lane. The sound travelling distance from car A to car B is shorter in the air. Car A sounds its horn and car B hears the sound. Next, car A is behind and sounds its horn again. Car B will found that the frequency of the sound from car A does not change even though the travelling distance changed.

风的影响没有产生在任何方向上的声音频率变化。声音的实验不能证明空气的存在。同样道理，Michelson–Morley 不能证明以太的存在。

There is no sound frequency change on any direction due to the wind. The sound experiment cannot prove the existence of air. For the same reason, Michelson–Morley experiment cannot prove the existence of aether.

上述实验其实是一个简单的多普勒效应实验。它告诉我们，由于多普勒效应，通过空气中声音的行驶距离变化不会导致声音的波长变化。

The above experiment is in fact a simple Doppler Effect experiment. It tells us that the due to Doppler Effect, the sound's travelling distance change through air does not cause sound's wavelength change.

在 Michelson–Morley 实验中，光发射器和接收器的相对速度为零。因为波长的变化只能存在发送方和接收方时移开或靠拢时产生，该实验不应该有任何波长的变化。

In the Michelson–Morley experiment, light emitter and receiver's relative speed is zero. There should be no wavelength change expected since wavelength change can only exists when sender and receiver move apart or moving closer together.

许多其他的相对论有关的实验使用的是波长测量的单向或双向的光速。他们是无效的，应该加以忽略。

Many other Relativity related experiments are using wavelength to measure the one-way or two-way speed of light as well. They are not valid either and should be disregarded.

激光测距仪测量工具可以测量双向光速。实验结果表明，双向光速在任何方向上是相同的。因此，SR 的时间空间的收缩是对的。不过，这并不证明，SR 单程速度的假设是正确的。

The two-way light speed can be measured with Laser Distance Measurement tools. The experiment results indicated that two-way speed of light is the same in any directions. Therefore, SR's time space contractions are real. But it does not prove that one-way speed assumption in SR is correct.

10.22.2　One-way Light Speed Experiment

有一对一公里长的光缆，沿直线连接三个站点。还有一个光缆连接两端的站点。中间站 C 点有一台计算机，一个计算机时钟和两个发射器。在两端的每个站点 A 和 B 都有一个原子钟，一个发射器，一个接收器和一台计算机。

There is a pair of one km long fiber cables connecting three sites along a straight line. There is one more cable that connects sites at the two ends. The site C in the middle has a computer, a computer clock and two emitters. Each site at the other two ends, A and B, has an atomic clock, an emitter, a receiver and a computer.

中间站点 C 在完全相同发送时钟纳秒值到 A 和 B。当 A 和 B 从 C 得到纳秒值，它们将自己的时钟设置成得到的纳秒值。然后 A 和 B 通过它们之间的 2 公里长的光纤电缆互相转发 C 的信息。

The middle site C sends its clock reading in nanoseconds at exactly the same time to A and B. When A and B get the message from C, they set their clock with the time received from C. Then A and B forward C's message to each other via 2km long fiber cable between them

从 C 到 A 要用 q 纳秒传送信息，而从 C 到 B 要用 p 纳秒。当 C 在 0ns 发出信息，A 收到信息把时钟设置为 0 而使时钟慢了 q 纳秒，同样 B 时钟慢了 p 纳秒，时钟设置为 0。当 A 从 B 得到信息，从 B 到 A 单程要用 2q 纳秒，A 时钟读数为 -q+2q= q。

It takes q to send message from C to A and p from C to B. When C emits message at 0ns, A sets its clock to 0 on receiving the message and introduce -q clock drift while B gets –p clock drift. When A gets message from B, light take 2q to travel one-way from B to A, the A's clock reading is -q+2q = q.

当 B 从 A 接收到该信息时，其时钟读数为 q。

When B receives the message from A, ts clock reading is p.

这两个时钟的读数之间的差是：

The difference between the two clock readings is:

p-q

下面测试参数的预测结果：

Predicted result for testing based on parameters:

t = 3335.640952ns, distance = 1000m and v = 200km/s:

从 B 到 A 的时间：

Elapse time from B to A:

q = 3331.1874 ns

从 A 到 B 的时间

Elapse time from A to B:

p = 3340.0945 ns

p-q = 8.9 ns

有必要在一天的不同时间，重复上述测试。

It is necessary to repeat the above test at different time of the day.

在一天的不同时间的预计的测试结果：

Predicted test results at different time of the day:

Time	p	q
12PM	3340.0945	3331.1874
3AM	3338.7865	3332.4954
6AM	3335.640952	3335.640952
9AM	3332.4954	3338.7865
12AM	3331.1874	3340.0945

光的最高传播方向是地球在以太中的移动方向。

The direction with maximum one-way light travelling time is the earth travelling direction in aether.

10.22.3　Spacetime Distortion

A 和 B 两个平行镜面以速度 v 相对螺旋网格移动：

There are two moving parallel mirror A and B with speed v relative to Torque Grids:

$v*t = B_0B_1 = B_1B_2 = A_0A_1 = A_1A_2$

在螺旋参考系中，光子从 A_0 到 B_1，然后移动到 A_2。

In the Torque reference, photon moves from A_0 to B_1, then to A_2.

在移动镜 A 看来，A_0，A_1 和 A_2 是镜子 A 上相同的点。在移动镜 B 看来，B_0，B_1 和 B_2 是镜子 B 上相同的点。光子在螺旋参考系中不同的点之间的反射；它却是在移动参考系的两个点之间反射。

From the moving mirror A stand point of the view, A_0, A_1 and A_2 are same spot on mirror A. From the moving mirror B stand point of the view, B_0, B_1 and E_2 are same spot on mirror B. The photon is bounced back and forth among different points in Torque reference; it is bounced back and forth between two points in the moving reference.

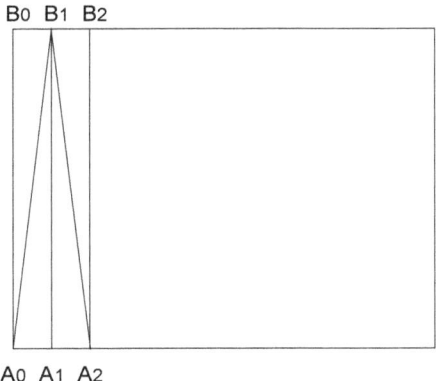

移动参考系用同样的光速测量的空间和时间。因此，它具有相同的光速，并拥有自己的时间和空间的测量值 l_m 和 t_m。

The moving reference uses the same light speed to measure space and time. Therefore, it has same light speed as Torque reference and has its own time and space measurements l_m and t_m.

$tc = A_0B_0$

$t_mc = A_0B_1$

$t_mv = A_0A_1$

$(tc)^2 + (t_mv)^2 = (t_mc)^2$

Or,

$$t_m = \frac{t}{\sqrt{1 - \dfrac{v^2}{c^2}}}$$

由于 $l_m = t_mc$ 和 $l = tc$:

As $l_m = t_mc$ and $l = tc$:

$$l_m = \frac{l}{\sqrt{1 - \dfrac{v^2}{c^2}}}$$

$$d = \sqrt{1 - \frac{v^2}{c^2}}$$

光可镜子沿运动方向或者逆镜子运动方向移动。

The light can travel in the direction of the mirrors' movement or opposite the direction of the mirrors' movement.

当光沿镜子移动方向移动:

When light travels along the moving direction:

$t_1 c = d(t_{m1} c + t_{m1} v)$

当光逆镜子移动方向移动：

When light travels opposite the mirrors' moving direction:

$t_2 c = d(t_{m2} c - t_{m2} v)$

$t_1 + t_2 = d(t_{m1} + t_{m2})$

To simply the above,

$t_1 + t_2 = 2t$

$t_{m1} + t_{m2} = 2 t_m$

$$t_m = \frac{t}{\sqrt{1 - \dfrac{v^2}{c^2}}}$$

光的移动时间是双向的移动时间。这意味着光的单向移动时间是不同的。

The travelling time of light is two-way travelling time. It implies that the light's one-way travelling time is different when travelling parallel to the moving direction.

空间/时间膨胀是发生在所有方向上。

The space/time expansion occurs in all directions with same factor d.

根据普朗克公式，$E = hc/L$。在移动参考系，当波长为 1 时，螺旋参考系中的波长是：

According to Planck equation, E=hc/L. In the moving reference, when the wavelength is 1, the Torque reference wavelength is:

$$\sqrt{1 - \frac{v^2}{c^2}}$$

Or,

$$E_{moving} = \frac{E_{torque}}{\sqrt{1 - \frac{v^2}{c^2}}} \qquad (10\text{-}5)$$

重复相对论的推导过程将得到：

Repeating relativity's process will get:

$$E = mC^2$$

10.23 Particle Waves and Torque Distortions

当粒子移动时，因螺旋波与粒子共振而尾随粒子移动。让我们来研究两个粒子 A 和 B。

When a particle is moving, the Torque waves initiated from moving particle are moving with particle as they resonance with the particle. Let's study two particles, A and B.

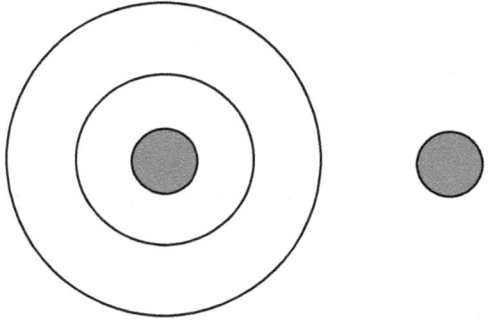

Figure 10-2

假定粒子 A 向 B 移动速度为 V，波长为 L，移动距离 L 所花费的时间为 T。

Assume that particle A is moving towards B at speed V, the wavelength is L, and the time it takes to travel distance L is T.

$$T = L/V$$

当 A 的螺旋波波周期的开始点碰到了 B，就开始了从 B 粒子反弹粒子波的周期。当 A 的螺旋波波周期的结束点碰到了 B，就结束了从 B 粒子反弹粒子波的周期。新的反弹粒子波从 B 粒子开始向外以光的速度传播。

When the beginning of the wave cycle for A hits B, the bounced back wave from B starts the particle wave cycle, when end of the wave cycle of A hits B, it ends the particle wave cycle. The new particle waves are the bounced back waves surrounding particle B traveling outward at the speed of light.

假设粒子波的波长为 λ。

Assume that the wavelength of the particle wave is λ.

$$\lambda = Tc = (L/V)c$$

从普朗克公式 E= hv，可得出：

From the Plank equation E=hv， we have:

$$E = h(c/L)$$

$$L = hc/E = hc/mc^2 = h/mc$$

$$\lambda = (L/V)c = \frac{h/mc}{V}c$$
$$mV = h/\lambda = h\upsilon \tag{10-6}$$

10.24 Light in Different Media

光子统过介质时，粒子波有同光线一致也有垂直的。这些垂直的波加大了波行走的距离，看上去光子变慢了。介质中的粒子对每一螺旋网格形变产生相同强度的垂直波，波长越短，垂直波越多，波行走的距离越长。所以，波长较短的光子在同一种介质中速度较慢。

When Photons pass through a media, the particle waves can be in same travelling direction with photon, or vertical to the travelling direction. The vertical waves increase the travelling distance and photon seem to be slower. The particles in the media create vertical waves with same strength per wave. A photon with shorter wavelength has more waves overall hence gets more vertical waves, and longer travelling distance. Therefore, the photon with shorter wavelength tends to be slower in the same media.

10.25 3D Particle Wave Strength

Suppose there is a sphere surrounding a particle. The total Torque Distortion of the particle on the Torque Grid remains constant regardless of the size of the sphere:

$$\oiint A dS = Km$$

其中：

Where:

K 是常数

K is a constant

m 是质量

m is the mass

对于一个半径为 r 的球体：

For sphere with a radius of r:

$$4\pi r^2 A = Km$$
$$A = Km / 4\pi r^2$$

上面的等式显明，随着距离的增加，单位面积的螺旋波强度会越来越小。

The above equation shows that the Torque Wave strength will be smaller and smaller per unit area as the distance increases.

同样，当距增加时，单位面积的粒子波的强度将越来越小。在双缝实验中，距离越远，干涉性越差。

For the same reason, the particle wave strength will be smaller and smaller per unit area as the distance increases. In the Two Slits experiment, the further the distance it gets, the less is the wave interference.

10.26 The Particle Wave in an Atom

因质量差的关系，原子核的运动可忽略不计。在原子中的运动电子的粒子波在原子核周围振动。

Due to the mass differences, the movement of nuclear in an atom can be ignored. The moving electron's particle waves are continuously vibrating around the nucleus.

在稳定的原子中，电子运动与其粒子波共振。

In a stable atom, the movements of electron are in resonance with its own particle waves.

图 6-3 显示一个电子的可能的移动方式。对于任何电子在极坐标中，主要有三个可能的转动，r，θ 和 φ：

Figure 6-3 displays the possible movements of an electron. For any electron in the polar coordinate, there are three main possible movements, r, θ and φ:

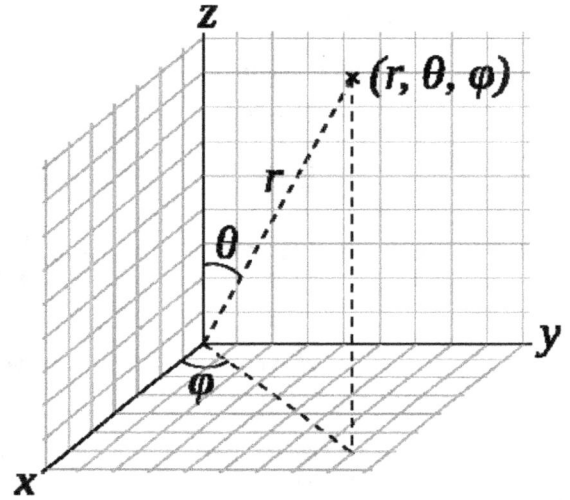

Figure 10-3

运动的三个方向与原子核作用形成了三个独立波。沿 r 方向的频率为 n，沿θ方向的频率为 l，沿φ方向的频率为 m_l。

The movement in three directions interacts with the nuclear and form three independent waves. The frequency along the r direction is n, along θ direction is l and along φ is m_l

n 波是主波：

The n wave is the main wave:

n=1,2,3,…

沿着θ旋转时，只有一个方向，因为这两个方向是没有区别，l = n-1，n-2，… ，0。

When circling along θ, it only takes one direction, since the two directions are no difference, therefore, l =n-1, n-2,…, 0.

请注意，l 小于 n。如果 L= N，它会作圆周运动，并不能保持稳定，如果 L> N，那么，L 就成了 n。l 不能是负的，因为为 l 的方向已经没有任何意义。

Please note that l is less than n. If l = n, it will make circular movements and cannot be stable. If l >n, then, l becomes n. l cannot be negative, since the direction has no meaning for l.

当 n 和 l 的方向决定后，m$_l$有两种可能的螺旋方向。m$_l$= l, l-1, l-2, 0, -1, -2, ...-l。正表示 Z 捻动和负代表 S 捻动。

Once the direction of n and l is decided, there are two possible twist direction for m$_l$. m$_l$= l, l-1, l-2, 0, -1, -2,...-l. The positive sign represents Z Twist and negative represents S Twist.

10.27 Charged Field Equation

在图 7-1 中的电子的外部螺旋场有松螺旋。

In Figure 7-1, the electron external torque field has a loosening torque.

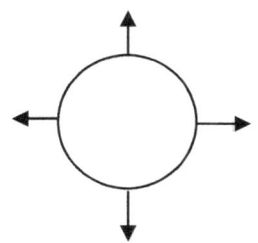

Figure 10-1

假设电荷的周围有一个球体。不论球体的大小如何，总的螺旋力不变。

Imagine that there is a sphere surrounding the charge. All the torque forces remain constant regardless of the size of the sphere.

$$\oiint EdS = Ch\arg e\,Torgue$$

其中：

Where:

E 是球体上的螺旋（电荷）力。

E is the Torque (charged) force on the sphere.

S 是球体的面积。

S is the area of the sphere.

假设球体的半径为 r：

Assume that the sphere has a radius of r:

$$4\pi r^2 E = e$$

$$E = \frac{e}{4\pi r^2}$$

10.28 Torque Interaction and Electronic Field

以下是两个带正电荷的粒子：

The follow are two positively charged particles:

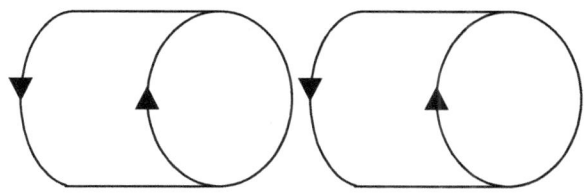

Figure 10-1

因为运动方向相反，两个电荷之间的螺旋运动相互抵消了。两个电荷远端的螺旋运动增强了。其结果是，两个带正电的粒子远离彼此。

The Torque movements between two charges cancel out since the movements are in opposite d rections. The farther ends torque movements for both charges are enhanced. As the result, the two positive charged particles move away from one another.

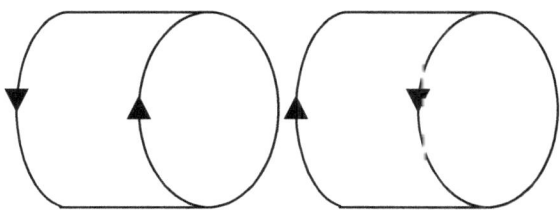

Figure 10-2

在上面的图中，有一个带正电荷的和一个带负电的粒子。两个粒子间的螺旋运动得到加强。两个粒子相互吸引。

In Figure 10-2, there is one positively charged and one negatively charged particle. The Torque movements between the two particles are enhanced. The two particles attract each other.

两个粒子的作用力的大小等于螺旋强度乘粒子的电荷。

The interaction force of the two particles equals to the torque strength times the particle charge.

$$F = A\frac{e_1}{4\pi r^2}e_2 = A\frac{e_1 e_2}{4\pi r^2} = \frac{e_1 e_2}{4\pi\varepsilon_0 r^2}$$

上面的方程是库仑定律。

The above equation is the Coulomb's Law.

10.29 Electron Positron Annihilation

下图从左至右是湮没过程的模型：

The following figure is a model of Annihilation process from left to right:

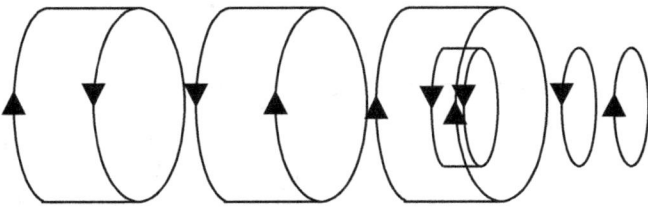

Figure 10-3

首先，两个电子，一个正电子和一个负电子，被对方所吸引。这两个电子之间的运动形成一个新的光子。新的光子移动到右侧。其他两个外侧的环形运动形成移动到左侧的第二个光子。

First, two electrons, one positron and one electron, are attracted by each other. The movement between the two electrons forms a new photon. The new photon moves out to the right. The other two outer circular movements create second photon which moves out to the left.

10.30 Charged Particle Circular Movement

当带电粒子作圆周运动，螺旋力矢量开始扭曲。磁场就是螺旋力矢量的扭曲场。

When the charge particles are circling, the Torque force vectors start to twist. The magnetic field is the electron vectors' twisting field.

Figure 10-1

扭曲的螺旋线的密度（假定每条螺旋线有相同的强度）和扭转速度的决定磁场的强度。下图研究具有以下特征的带正电的圆周运动：

The density of the twisted lines (assume that each twisted line has the same strength) and the twisting speed determines the strength of the magnetic field. The next diagram studies a circling positively charged object with the following characteristics:

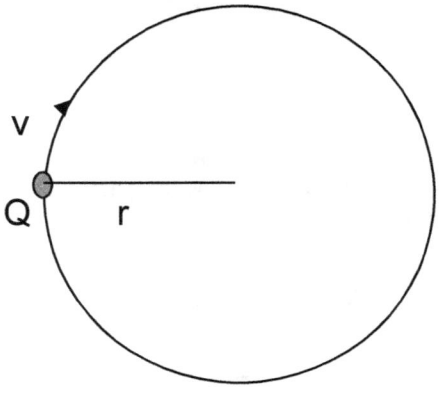

Figure 10-2

v 是速度

v is the speed

Q 是的电荷量

Q is the amount of charge

r 是半径

r is the radius

B 是磁场强度

B is the strength of the magnetic field

$$B=Qv/r$$

10.31 Ampere's circuital law

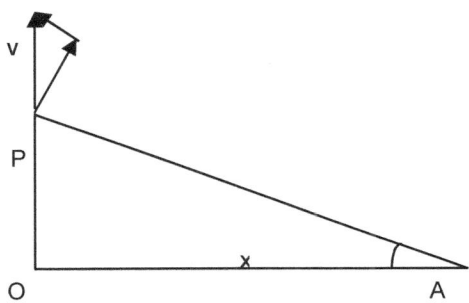

Figure 10-3

在 P 点的带电荷 Q 的粒子以速度 v，沿 OP 运动在 A 点产深生沿 x 方向的磁场 AP=r, OA=x, OP=y）：

A particle with charge Q, speed v at point P moves along OP has following magnetic field on A point (AP=r, OA=x, OP=y):

$$Qv(1/\sqrt{x^2+y^2})(1/\sqrt{x^2+y^2}) = Qv/(x^2+y^2)$$

假设电线 OP 的长度是无限的。OP 的单位长度有 Q 电荷，而 v= 1 / t。磁场的强度是：

Assume that the length of the wire OP is infinite. A unit length on OP has Q charge, then v=1/t. The strength of magnetic field is:

$$\int_{-\infty}^{+\infty} \frac{Qv}{x^2+y^2}\,dy = \frac{1}{x}\int_{-\infty}^{+\infty} \frac{Qv}{1+\left(\dfrac{y}{x}\right)^2}\,d\left(\frac{y}{x}\right)$$

Assume y/x =tan(t):

$$\frac{1}{x}\int_{-\pi/2}^{+\pi/2}\frac{Qv}{1+(\tan(t))^2}d(\tan(t))=\frac{1}{x}\int_{-\pi/2}^{+\pi/2}Qvdt=\frac{\pi Qv}{x}$$

由于 v=1 /t，电流 I=Q/t：

Since v=1/t, and electric current I=Q/t:

$$B=\pi Qv/x=(\pi Q/x)/t=\pi I/x$$

或：

Or:

$$2\pi xB=\pi^2 I$$

上面的方程的一般形式为：

The above equation can be generalized as:

$$\oint Bdl=\mu_0\oiint_s J_f dS \qquad (10\text{-}7)$$

这个公式是安培环路定律。

This equation is the Ampère's circuital law.

10.32 Light Speed

由于螺旋运动运载光子，在真空中，光子与螺旋运动速度相同。

Since Torque movements carry the photon, a photon has the same speed in the vacuum as Torque movements.

11 Authors' Biography

我毕业于中华人民共和国上海交通大学。我在大学的本科的专业是材料科学，硕士学位是计算机科学。物理和数学是我最喜欢的课程。我来到美国继续攻读计算机科学，在底特律韦恩大学获得硕士学位。作为一个电脑工程师，我在美国做了约 24 年工作。

I graduated from Shanghai Jiao Tong University in the People's Republic of China. I majored in Materials Science with a B.S. degree, and a M.S. degree in Computer Science. Physics and mathematics were my favorite classes. I came to the United States to pursue further study in Computer Science, and I got another Master Degree in Wayne State University in Detroit. I have worked as a computer engineer in the United States for about twenty four years.

在大学里，我对夸克理论，广义相对论，尤其是与黑洞有关的猜测有疑问。因我不能接受这些无法解释的理论，而决定去寻找答案。我花了大多数时间研究夸克理论，最终发现夸克理论是完全错误的。我创建了螺旋模型来解释一切我想解释的东西。这本书搜集了我多数的发现。

In college, I had doubts with quark theory, general relativity theory, especially with black hole related speculations. I could not accept these unexplainable theories and decided to find the answers. I studied Quark theories for most of my life and eventually found that the theory is completely wrong. Then, I created the Torque model to explain everything I wanted to explain. This book collects most of my discoveries.

曹海明为螺旋理论的各个方面提供了重要的见解，其中包括斐波那契数列和重力理论等。他还为这本书做了最后的编辑。

Henry Cao provided vital insights to every aspect of the Torque theory such as the Fibonacci series and Gravity theory. He also did the final editing for this book.

www.ingramcontent.com/pod-product-compliance
Lightning Source LLC
Chambersburg PA
CBHW071234170526
45165CB00003B/1096